*The twentieth century has seen biology come of age as a conceptual and quantitative science. Biochemistry, cytology, and genetics have been unified into a common framework at the molecular level. However, cellular activity and development are regulated not by the interplay of molecules alone, but by interactions of molecules organized in complex arrays, subunits, and organelles. Emphasis on organization is, therefore, of increasing importance.*

*So it is too, at the other end of the scale. Organismic and population biology are developing new rigor in such established and emerging disciplines as ecology, evolution, and ethology, but again the accent is on interactions between individuals, populations, and societies. Advances in comparative biochemistry and physiology have given new impetus to studies of animal and plant diversity. Microbiology has matured, with the world of viruses and procaryotes assuming a major position. New connections are being forged with other disciplines outside biology—chemistry, physics, mathematics, geology, anthropology, and psychology provide us with new theories and experimental tools while at the same time are themselves being enriched by the biologists' new insights into the world of life. The need to preserve a habitable environment for future generations should encourage increasing collaboration between diverse disciplines.*

*The purpose of the Modern Biology Series is to introduce the college biology student—as well as the gifted secondary student and all interested readers—both to the concepts unifying the fields within biology and to the diversity that makes each field unique.*

*Since the series is open-ended, it will provide a greater number and variety of topics than can be accommodated in many introductory courses. It remains the responsibility of the instructor to make his selection, to arrange it in a logical order, and to develop a framework into which the individual units can best be fitted.*

*New titles will be added to the present list as new fields emerge, existing fields advance, and new authors of ability and talent appear. Only thus, we feel, can we keep pace with the explosion of knowledge in Modern Biology.*

*James D. Ebert*
*Ariel G. Loewy*
*Richard S. Miller*
*Howard A. Schneiderman*

**Modern Biology Series**



**Published Titles**

| | |
|---|---|
| *Burnett-Eisner* | **Animal Adaptation** |
| *Delevoryas* | **Plant Diversification** |
| *Fingerman* | **Animal Diversity** |
| *Levine* | **Genetics,** *second edition* |
| *Odum* | **Ecology** |
| *Ray* | **The Living Plant** |
| *Savage* | **Evolution,** *second edition* |
| *Sistrom* | **Microbial Life,** *second edition* |
| *Van der Kloot* | **Behavior** |

**Forthcoming Titles**

| | |
|---|---|
| *Ebert-Sussex* | **Interacting Systems in Development,** *second edition* |
| *Ehrenfeld* | **Biological Conservation** |
| *Gillespie* | **Cell Function** |
| *Novikoff-Holtzman* | **Cells and Organelles** |

# *Animal Structure and Function*

**second edition**

***Donald R. Griffin***
*Rockefeller University*

***Alvin Novick***
*Yale University*

*Holt, Rinehart and Winston, Inc.*
*New York Chicago San Francisco*
*Atlanta Dallas Montreal*
*Toronto London Sydney*

*Library of Congress Catalog Card Number: 70–77810*
ISBN: 0-03-077505-1
*Printed in the United States of America*
34567890 006 9876543

*Text and cover design by Margaret O. Tsao*
*Illustrations by Gaetano di Palma*

# *Preface*

Our prevailing theme in writing this book (and in our teaching) has been the biology of the whole animal. The exploration of, enlargement on, illustrations of, and digressions from the theme have been along a few consciously chosen paths. Classically, the organism has been described anatomically. We have not wished to overlook this wealth of beautiful material and we have included a number of examples, in depth, of anatomy and of comparative anatomy. Primarily, however, we have tried to examine the functions of animal organ systems and their regulatory and integrative mechanisms. Animal function, of course, goes hand-in-hand with anatomy. The two views are often and ideally inseparable. Generally, therefore, we have discussed these as a unit.

Studies of organismal physiology have entered an exciting new phase in the last decade or two. The subfields of comparative and environmental physiology have matured and have become effective contributors to our general understanding of organisms. One major aim of this revised edition has been to incorporate a selection of illustrative examples taken from these fields into our discussions. Any reasonable examination of water or ion regulation in animals, or of respiration, for example, must be based on comparisons of organisms from different phyla (or different genera or species) living in different adaptive niches and being buffeted by different environmental conditions.

Animal behavior has also become an important field of investigation in recent years. Advances here and in sensory physiology have fleshed out the bones of neurophysiology and of endocrinology so as to have made it appealing to us to greatly enlarge the chapters on reproduction and on the coordination of function.

Basically, we have aimed at giving students a comprehensive view of what an organism is and what it does. We have given little attention to molecular or cellular levels of organization but, on the other hand, we have not ignored these levels when their treatment fitted our problem. Most of our interest and attention has been delegated to those realms—nutrition, metabolism, locomotion, circulation, chemical regulation, reproduction, and integration of function—which seem to have special organismal meaning. In each chapter we have tried to combine a basic description of organizing principles with a vivid array of comparative or environmental examples.

We are grateful to many people who have helped create this book, but we wish, most of all, to express our appreciation for the contribution made by our enthusiastic, probing, and even trying students who have listened to us, judged us, taught us, and made our occupation meaningful.

*New Haven, Connecticut* — *A.N.*
*New York City* — *D.R.G.*
*December 1969*

# *Contents*

*chapter* **1**

# *The Significance of Animals in Modern Science*

Out of the whole universe, nothing we know or even seriously imagine can compare with the intricate machinery that makes up living animals. The physical sciences restrict their scope to systems lending themselves to simple explanations. The proper study of animals and plants, however, calls for a higher order of analysis and understanding. Biology is intrinsically more complex than physics or chemistry. Our own bodies and our ways of life are largely the consequence of its orderly complexities.

Men have long recognized that animals are our distant relatives. Yet just a century ago this was a radical doctrine that generated an emotional cyclone about the heads of Darwin and his fellow thinkers when they called for a readjustment of basic beliefs about our relationship to the rest of the universe. No other idea since the Copernican revolution has so profoundly altered human thinking as the biological concept of evolution; this subject is considered in *Evolution,* another book in this series. We shall take it for granted here that living animals and men are descended from different, usually simpler, animals that lived in the past and, furthermore, that most of these lines of descent can be traced back in reasonable detail through a fossil record of some hundreds of millions of years.

One striking attribute of animals is their diversity; the patient labors of several generations of biologists have left us with an embarrassment of factual riches about the million-odd forms of animal life with which we share this earth. Practically all have been given names, and almost every part of any given sort of animal has also been identified, so one biologist may know what another is talking about. This stockpile of information about animals staggers the imagination and sometimes stuns the intellect. Fortunately, many important and exciting facts have recently been learned about the efficient design of animal bodies and the manner in which their many parts work together in functional harmony.

## *CELLS, MICROBES, PLANTS AND ANIMALS*

Close examination of any animal shows that it is built up from microscopic units of organized life called *cells.* These are usually a few microns in length or width (1 micron or $\mu = 10^{-6}$ meter or about 1/25,000 inch) and are surrounded by cell membranes so thin that their presence can barely be detected with the best microscopes employing visible light. So many of the basic phenomena of life occur in the viscous material called protoplasm that is bounded by cell membranes that an entire book, *Cell Structure and Function* by Loewy and Siekevitz (see Further Reading at the end of this chapter), is devoted wholly to the fundamental processes that occur within the confines of single cells. Many living organisms consist of only one cell (viruses consist of only part of a cell); these are usually known as microbes and are dealt with in detail in *Microbial Life,* a book in this series. But those organisms we ordinarily call animals are composed of thousands—or, more often, millions—of cells specialized into dozens or hundreds of distinct types. Because of the specializations, very few of these cells are capable of living independently outside the organized body of which they form a part. Yet when functioning together, they add up to an animal that does things that would be impossible for a population of microbes.

Plants, of course, are also multicellular, living organisms. They are distinguished from animals, in general, by their relative lack of movement and by their ability to synthesize food materials from simple molecules by trapping the energy of sunlight in the process called photosynthesis. (Plants and microbes that can carry out photosynthesis are called *autotrophic*, in contrast with the typical animal, which is designated *heterotrophic* because it must obtain the energy it needs by consuming, as food, molecules that have been synthesized by plants or other living organisms.) There are notable exceptions to this general rule. Many animals live part of their life in a relatively inactive state, often attached to a substrate in the ocean; examples are mussels and clams, or the corals, which secrete around their bodies hard, protective shells that pile up by the millions to form coral reefs. Even among these animals, appendages usually move to and fro, and gametes and larval stages of the life cycle are likely to be very active. Flowers may open and close their petals every day, leaves rise and fall, stems twist, and many of the microscopic aquatic plants called algae swim actively at some stage in their lives. Even among higher plants, the gametes are often active. Furthermore, there are plants, such as the fungi, that are incapable of photosynthesis and are quite as heterotrophic as a codfish or a cow. Complicating matters even further, certain corals and related animals have intimate associations with single-celled algae. The algal cells are scattered among those of the coral and carry out photosynthesis for the mutual benefit of themselves and the surrounding animal (a relationship called *symbiosis*), and thus the combination becomes an autotrophic entity.

Among microbes the distinction between animal-like, heterotrophic cells and plantlike, autotrophic ones becomes even more difficult to follow. In some cases, the very same cell changes during its lifetime from being autotrophic and motionless to swimming actively and ceasing to carry out photosynthesis; or, it may be mobile and autotrophic at the same time. When biology was dominated by the effort to classify living organisms, zoologists who studied animals claimed the mobile microbes to be members of the animal kingdom and botanists classified the photosynthetic ones as plants. These criteria led to some entertaining jurisdictional disputes when the same microbe changed its way of life. Modern biologists, however, place emphasis on the distinction between microbes and multicellular organisms rather than on a jurisdictional dispute. Many significant processes are regulated at cell membranes or occur within single cells in both groups. Quite different phenomena present themselves, however, when we consider how large numbers of cells are organized into plants or animals. In plants, most of the "sociology of cells" serves to make them more efficient in a sessile, autotrophic way of life. In animals, on the other hand, cells are so assembled and arranged that the resulting organism moves actively to obtain its food and to carry out other independent activities. Microbes and plants synthesize complex molecules and grow, but animals *do* things. What they do and how they do it depends very largely on how they are constructed, what organs make up their bodies, and how their multitudes of cells are joined together into a stable, effective community of cooperating parts. The central purpose of this book is to outline as clearly as possible some of the ways in which cells are functionally organized into animals.

## *THE ORGAN SYSTEMS OF ANIMALS*

To analyze the workings of even one small part of an animal is surprisingly difficult because so many events are taking place concurrently within microscopic dimensions. We can simplify this task by considering the important functions one at a time. This approach is facilitated by the fact that cells and organs are more or less specialized in an efficient division of labor to perform each of the several functions that are essential for every sort of animal. An organism can be viewed as a collection of individual cells. Groups of similar cells such as those making up cartilage or muscle are referred to as a *tissue*. Several tissues are usually combined in an organized fashion to make up an *organ* (such as a kidney, an eye, or a bone). Several organs usually function together to make up an *organ system*, such as the digestive system or the skeletal system. Systems taken together constitute an *organism*. The nature of the systems and their complexity of function will vary, of course, with the life stage (egg, larva, adult) of the organism.

We can distinguish eight major functional systems: framework, digestion, respiration, mobility, internal transport, regulation of chemical composition, reproduction, and integrational regulation of function. Growth, an important function, of course, is considered in *Development* in this series.

Animal bodies are held together in a recognizable and more or less specific shape that provides a framework for all other operations and processes. The outermost layers always form a skin of varying thickness and strength; in many animals there is also a more rigid supporting system, or skeleton, either inside the body or associated with the skin. The nature of these supporting systems varies widely according to the needs of the animal and its basic organization.

In addition to a supporting system, an animal needs some sort of fuel for its body machinery—in other words, food. This it must obtain from the outside world by some sort of positive action, such as the search for, pursuit of, and ingestion of prey, after which the food must by processed in many ways before being taken into the body proper. This processing, called digestion, is the mechanical and chemical breaking up of raw food into molecules directly usable for the production of energy and for repair and growth.

Along with food molecules, most animals require a supply of oxygen. The principal end products of the whole process of oxidation of food are carbon dioxide and water in addition to energy. The obtaining of oxygen and the disposal of excess carbon dioxide are the chief functions of respiratory systems, usually gills in aquatic animals and lungs or tracheae in terrestrial animals.

Animals characteristically use much of their supply of energy to move about. Both the relative motion of various parts of the animal and the locomotion of the whole animal require muscles, the operations of which can be understood only in terms of their organization at the visible level as well as the organization of the contractile protein molecules within the specialized muscle cells.

Most animals of any size require a specialized circulatory system to distribute many substances among their various organ systems. Blood vessels and hearts are often critically important organs whose failure causes rapid death.

The maintenance of life requires much more than the straightforward operations just listed, for living machinery is too complex to run without regulation and control. The chemical regulation of an animal's body fluids is especially important, since only the correct concentrations of salts and other small molecules will sustain life. Thus in the most highly organized animals, for efficient function, the ionic composition of the body must be regulated with great precision. This regulation of water and ions is carried out chiefly by organs, such as the kidneys, which may have some excretory functions as well.

Most organic machinery in an animal sooner or later wears out, sometimes causing death, but most animals, in nature, die from disease, predation, or accident. Life persists because new animals have usually been produced in the meantime. Reproduction requires a major share of the energies of mature animals. Yet the construction of replicas of the parents is not sufficient. Over the long course of evolution a more complicated process, sexual reproduction, has prevailed because it is better suited to perpetuate animal species over time spans far longer than individual lifetimes. The reasons for superiority of sexual reproduction over exact duplication may be deduced after studying *Genetics* and *Evolution* in this series. Reproductive systems are required to supply the specialized cells called gametes that, under favorable conditions, fuse to form the beginning of a new animal. Many animals also have elaborate organs to care for their developing young.

A bundle of organs is still not an animal. Even though certain of these organs, such as the kidneys, regulate the chemical composition of the body fluids, there remains the problem of regulating and coordinating a myriad of other processes as well as the various activities of muscles and other organs. Rapid, split-second adjustments are often required; others, though, are spread over hours or days and call for equally critical balancing. In response to these needs, a variety of mechanisms have evolved for regulating the contractions of muscles, the secretory activity of glands, and the rates at which other organs function. Some of this regulation is achieved by dispatching molecular messengers, called *hormones*, from one part of the body to another. Whole systems of endocrine glands are used to produce hormones; some have widespread effects, whereas others regulate the temporary or long-term actions of specific organs. The coordination of rapid movement is achieved by more specialized signals that travel over the surface of nerve cells; a distinct nervous system is present in most animals to provide for the thousands of such signals that must flow back and forth to keep the organism's many parts performing correctly and at the proper, coordinated rates. The nervous system, more than any other system, makes animals what they are—independent operators, organisms that do things on their own.

## FURTHER READING

Barrington, E. J. W., *Invertebrate Structure and Function.* London: Nelson, 1967.

Clements, A. N., *The Physiology of Mosquitoes.* New York: Pergamon, 1963.

Davson, H., *A Textbook of General Physiology*, 3d ed. Boston: Little, Brown, 1964.

Florey, E., *An Introduction to General and Comparative Animal Physiology.* Philadelphia: Saunders, 1966.

Gordon, M., *Animal Function: Principles and Adaptations.* New York: Macmillan, 1968.

Greenwood, P. H., *A History of Fishes* (revision of J. R. Norman). London: Benn, 1963.

Hoar, W., *General and Comparative Physiology.* Englewood Cliffs, N.J.: Prentice-Hall, 1966.

Loewy, A. G., and P. Siekevitz, *Cell Structure and Function*, 2d ed. New York: Holt, Rinehart and Winston, 1969.

Marshall, A. J. (ed.), *Biology and Comparative Physiology of Birds.* Vols. 1 and 2. New York: Academic Press, 1961.

Moore, J. A. (ed.), *Physiology of the Amphibia.* New York: Academic Press, 1964.

Morton, J. E., *Molluscs*, 3d ed. London: Hutchinson University Library, 1964.

Mountcastle, V. B., *Medical Physiology.* New York: Mosby, 1968.

Nicol, J. A. C., *The Biology of Marine Animals.* New York: Interscience, 1960.

Prosser, C. L., and F. A. Brown, *Comparative Animal Physiology*, 2d ed. Philadelphia: Saunders, 1962.

Ruch, T. C., and H. D. Patton, *Physiology and Biophysics*, 19th ed. Philadelphia: Saunders, 1965.

Scheer, B. T., *Animal Physiology*. New York: Wiley, 1965.

Wigglesworth, V. B., *Insect Physiology*, 6th ed. New York: Wiley, 1966.

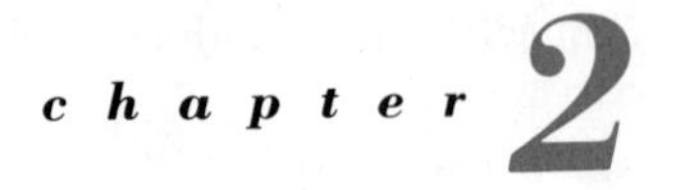

# chapter 2

# The Body Plans of Animals

For all their bewildering variety, animals are not infinitely diverse. We need to understand only a few major types of body plan to obtain a useful frame of reference into which their varying structures and functional properties can readily be fitted. This is not accidental. Once the concept of evolution is accepted with understanding, it makes excellent sense to think of large numbers of related kinds of animals as descendants of one ancestral population. This common ancestry explains much of the basic similarity in body plan that characterizes the principal groups or phyla of animals. Detailed descriptions of the major phyla and their probable relationships to one another may be found in textbooks of invertebrate zoology and in *Animal Diversity* in this series.

## *DIVISION OF LABOR WITHIN SINGLE CELLS*

Of all the body plans of animals, the superficially simplest is that of the animal-like microbes—namely, a single cell. These are often called Protozoa. In common with all cells, they have the following basic parts, starting from the outside and working inward: cell membrane, often including cilia or flagella; cytoplasm, containing mitochondria, endoplasmic reticulum, ribosomes, and other specialized components; and a nucleus, surrounded by a nuclear membrane and containing a nucleolus and chromosomes. For a thorough description of the basic structure of a living cell, see *Microbial Life* in this series and *Cell Structure and Function* by Loewy and Siekevitz (listed in Further Reading at the end of this chapter). Many small protozoans are no larger, and no more complex, than many of the relatively unspecialized cells found in the bodies of larger animals, yet they are still able to survive on their own in a variety of ways of life. A great many protozoans, however, are much larger and more complex than most cells of large animals. They often contain specialized structures, called *organelles*, that perform functions that, in large animals, are carried out by large groups of cells organized into organs or organ systems. The operation of these larger, complex cells apparently places more demands on the nucleus than is the case for a small cell, and many such protozoans have multiple or greatly enlarged nuclei. In the large and successful group of complex protozoans known as the ciliates, two kinds of nuclei are found; one is specialized for holding the chromosomes containing the substance deoxyribonucleic acid (abbreviated DNA), which carries the genetic "instructions" that will be passed on to daughter cells; the other contains many copies of the genetic instructions, in the form of DNA. These copies are used for the synthesis of a related substance, ribonucleic acid (RNA), which carries the genetic "instructions" into the cytoplasm, where they can be used for the synthesis of the proteins needed by the cell. The success of this group of protozoans suggests that it is advantageous for them to have two specialized kinds of nuclei, but we do not yet understand why this is so.

Protozoans often have skeletons. In some species of Protozoa, grains of sand or other tiny particles adhere to the sticky external surface of the cell membrane so that the cell comes to resemble a mobile croquette. Others have intricate internal skeletons, formed by the secretion of an interconnected series of rods and plates. These may be surrounded by soft and mobile protoplasm, or the plates may form a protective outer shell provided with a few holes through which temporary tentacles of protoplasm are projected for feeding. Some of these skeletons are formed primarily of calcium carbonate ($CaCO_3$), others of silicon dioxide ($SiO_2$), and some are made of still other materials. Often they form beautiful shapes and patterns.

Many protozoans contain organelles known as *food vacuoles*, in which food particles are digested, just as food is digested in the complex digestive systems of multicellular animals. In some protozoans (amoebae), the whole body surface is designed for the engulfment of food particles; in many other cases, a specialized region of the surface, known as the *cytostome*, is the organelle that functions as a mouth. Some ciliates have another specialized part of the cell membrane that serves as an anus, through which the undigested remains of food are ejected. Another organelle found in many protozoa is known as the *contractile vacuole*. Its function, which will be described more fully in Chapter 8, is to get rid of surplus water that has entered the cell; it is thus the functional equivalent, in part, of the kidney of higher animals.

Thus, even within a single protozoan cell, there is intracellular specialization—that is, the carrying out of a particular function by an organelle. Protozoa contain so many highly specialized organelles (some of which will be mentioned in later chapters), packed in such a small volume, that it is often simpler to analyze such functions as digestion in multicellular animals, where the division of labor is on a larger scale.

## *SIMPLE BUT EFFECTIVE COMMUNITIES OF CELLS*

Unicellular plants often form small aggregates, in which each cell functions almost exactly as it would alone. When multicellular animals are formed by the joining together of many cells, there are more stringent demands regarding cooperation among cells. However, teamwork does not seem to pay on any large scale until the "team" reaches a considerable size. It seems to be more efficient either to remain unicellular (which has advantages for survival—in the ease of dispersal and the minute size of the morsel a protozoan presents to predators, for example) or to form an animal with fairly large organs, each containing hundreds of cells. There are, of course, many other considerations related to size and unicellularity, such as surface-to-volume ratio and the volume of cytoplasm that a single nucleus can serve. To be sure, all multicellular animals begin life as a single cell and are necessarily obliged to go through a series of immature stages involving increasing numbers of cells. These developmental stages, however, may be rushed through or specially protected. There are, of course, also multicellular animals of very small size and often of very simple construction. These phyla, in general, are still poorly known physiologically.

## *THE COELENTERATE BODY PLAN*

There is one fairly simple body plan that is remarkably successful in numbers of species and individuals in spite of its simplicity.

Animals built according to this plan are, in essence, a lined sac, the inner lining consisting of a continuous layer of one type of cell and the outside covering consisting of a layer of cells suited for protection from the surrounding world. A closed sac would have little possibility of commerce with the world around it, so, in this efficient plan, an opening into the inner cavity serves as both mouth and anus. Food is taken in through this opening and undigested residues are eliminated by the same route. This body plan is characteristic of the phylum Coelenterata, which means saclike animals. To this phylum belong the jellyfish and corals along with many basically similar animals such as the fresh-water hydras and the sea anemones. (See Fig. 2-1.)

One side of the body of a coelenterate (the *oral* surface) contains the mouth and is surrounded by a ring of tentacles that are used to bring in food, and the opposite side (the *aboral* side) is sometimes specialized for attachment to a surface. However, there are no other specialized surfaces, so these animals tend to be symmetrical around a central axis—that is, they are *radially symmetrical*. Within the inner

***Fig. 2-1*** *Representative coelenterates.* (Left) *A branching, colonial hydroid.* (Right) *A medusa or jellyfish.*

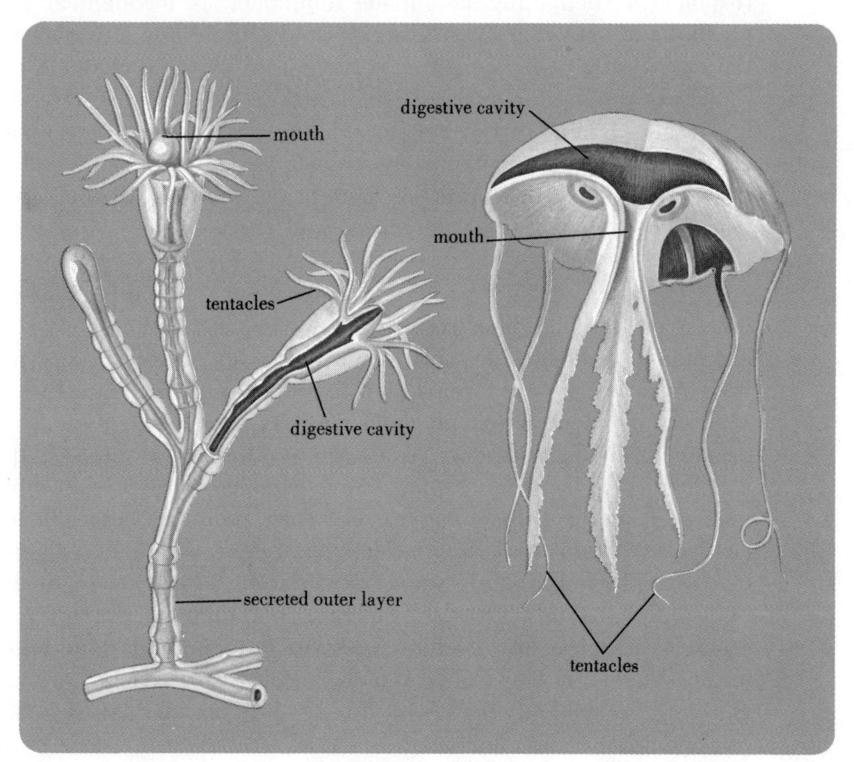

cavity formed by the saclike shape of the body, much of the food is digested. Digestive enzymes are liberated into this cavity, which for this reason may well be called the digestive cavity, although it serves other purposes as well. Some cells lining the digestive cavity take food into their cytoplasm much as protozoans do, but the organization of great numbers of cells into a large sac permits much larger prey to be taken and digested. Furthermore, this arrangement permits many cells to concentrate on other operations—for example, the movement of the tentacles. The amino acids and sugars resulting from the digestion of food within the digestive cavity are distributed throughout the ramifications of this cavity, which extend even into the tentacles. The division of labor is therefore a mutually effective one: cells of the tentacles operate to capture large prey, and many cells lining the digestive cavity are specialized to some extent for digesting food.

The cells of the inner and outer layers are specialized in many other ways. Some are muscle cells that move the tentacles, bend the main part of the body, or in times of danger contract both tentacles and body into a compact cylinder, with the digestive cavity reduced to negligible volume. There are long nerve cells that coordinate these muscular contractions and there are specialized sensory cells that respond to such different outside influences as mechanical contact, gravity, and light. There are also cells that secrete protective layers over the outer surface of many coelenterates; an extreme example of such a layer is the hard, calcareous shell of the corals. The inner and outer layers may be one or two cells in thickness and the two layers are not always in close proximity. In most coelenterates, a jellylike substance is secreted by the cells of one or both layers to fill out the body wall between them to a greater thickness, and occasional cells may be dispersed in this jelly.

The coelenterate body plan has almost unlimited possibilities for expansion. The digestive cavity and the surrounding body wall, for example, may be long or short. Furthermore, one, two, or many mouths, with tentacles, may be formed for a common, branching digestive cavity. This arrangement is extended, in many types of coelenterates, into a branching colony with dozens or even hundreds of mouths. With such a colony, one has difficulty deciding what constitutes the individual animal. Is it the single complex of mouth, tentacles, and adjacent digestive cavity or is it the whole colony? If the former, such a coelenterate exists as a sort of multiple Siamese twin; if the latter, there is no definite limit to the size to which the branching, plantlike body may eventually grow. Although many coelenterates are attached to some substrate for most of their lives, others such as jellyfish are active swimmers. Here the saclike body plan is modified to form an undulating, umbrella-like membrane surrounding the mouth. The jellyfish swims by waves of contraction of this umbrella. Despite their simple body plan and often plantlike appearance, however, the coelenterates are animals composed of hundreds of specialized cells with division of labor among them. But the number of different kinds of specialized cells is limited. Many of their activities are directed toward movement and a heterotrophic way of life devoted to the capture of other living organisms as food.

## THE FLATWORM BODY PLAN

For all their success in populating the seas and building coral reefs, coelenterates still remind us in many ways of plants. Another large group of animals that crawl and may also swim actively are the flatworms (phylum Platyhelminthes). The digestive system of this group distinguishes the flatworms from other wormlike animals; it has a single opening, as in the coelenterates, whereas all other major groups of animals have a tubular digestive tract beginning with a mouth for the intake of food and ending with an anus through which the undigested residues or feces are eliminated. Many flatworms are parasites and live inside the bodies of other animals, but we will concentrate here on free-living members of this and other groups of animals. Flatworms, unlike the radially symmetrical coelenterates, have a definite head and usually move head first; they also have a preferred orientation, with one surface up and the other down or pressed against the surface on which they are moving. The single opening of the digestive tract is on this lower or contact surface, which is usually called the *ventral* surface (by derivation from the Latin root meaning stomach). The opposite surface, usually uppermost, is called *dorsal*, meaning back. The head end of an animal is called the *anterior*; the opposite end, the *posterior*. This kind of shape, where the right half is approximately a mirror image of the left, and where the animal has a dorsal and ventral surface, is called *bilateral symmetry*. (See Fig. 2–2.) Active animals are usually bilaterally symmetrical. Sessile forms, which usually filter or gather food as it passes by, are often radially symmetrical.

Bilaterally symmetrical organization of the body is accompanied by a concentration of the nervous system into the anterior end, called *cephalization*. This concentration is what distinguishes the front end as the head. Sense organs, such as aggregates of light-sensitive cells that form simple eyes, are most likely to be found in the head, although in flatworms and many other animals some light-sensitive cells are found elsewhere over the surface of the body. They may even be found deep in the body if the tissues are transparent or translucent.

Although there is only one opening from the digestive cavity to the outside, only with difficulty can the body of a flatworm be conceived

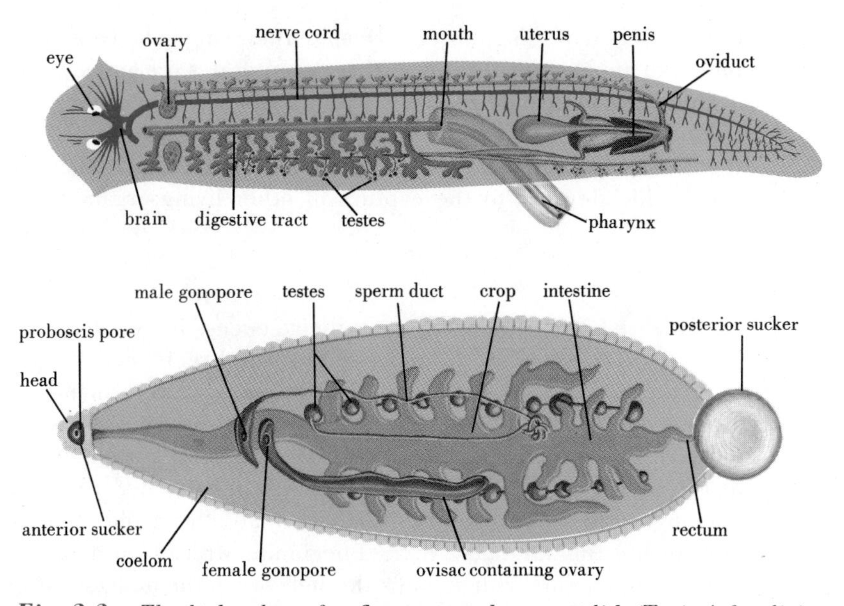


*Fig. 2-2 The body plan of a flatworm and an annelid.* (Top) *A free-living flatworm, showing the relative positions of the nervous, digestive, and reproductive systems.* (Bottom) *A leech, an annelid which feeds on blood. A fluid-filled cavity, the coelom, surrounds the gut and reproductive organs. The annelid nervous system appears in Fig. 10-3.*

of as a sac. The digestive cavity extends, with branches and ramifications, to almost all parts of the ribbon-shaped body and whole, complex organs are fitted between it and the outer skin. Aside from the nervous system, there are also specialized organs for reproduction and for regulation of body fluids. These organs will be discussed in later chapters. Here the division of labor is not merely between cells, as in the coelenterates; it is between organs.

## THE ROUNDWORM BODY PLAN

The roundworms constitute a large group of active, cylindrical, wormlike animals that differ from the flatworms in having a tubular digestive tract in which food moves from mouth to anus, undergoing digestion on the way. Many of the roundworms have a relatively large, fluid-filled cavity *between* the body wall and the digestive tract. The digestive tract, along with reproductive organs, is suspended in this fluid-filled cavity with some freedom to move relative to other parts of the body as the worm twists and swims or wriggles. (See Fig. 2-3.) These worms often have special mouth parts, with teeth, for grasping food; the digestive tract may be somewhat specialized along its length

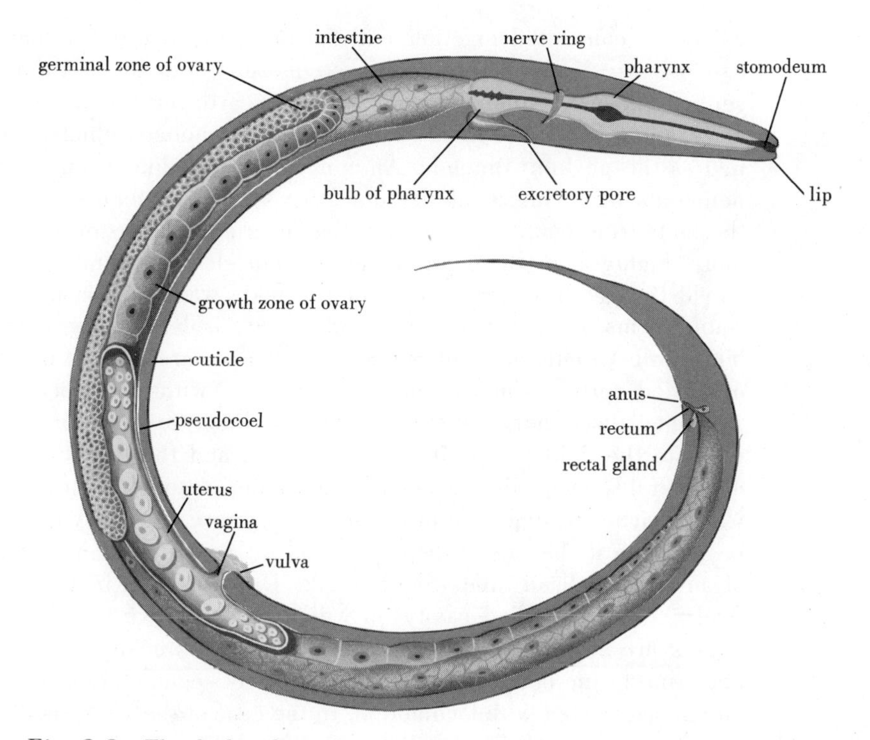


***Fig. 2-3*** *The body plan of a female nematode. A fluid-filled pseudocoel surrounds the digestive and reproductive systems, separating them from the body wall. The animal is not segmented.*

for the processing of food. See *Animal Adaptation* in this series for a perceptive analysis of the roundworm's way of life. The body cavity, distinctive from the digestive tract and not connected with the outside, serves many purposes including the circulation of its fluid contents in a way that serves to move materials from one end of the worm to the other.

Although there are quite a few different groups of animals constructed according to the roundworm body plan, most of them are relatively inconspicuous. The best known are the nematodes, a classification that includes a variety of relatively large forms (living within organs such as the lungs or digestive tracts of other animals) and small forms (which are abundant and important inhabitants of soils).

## THE ANNELID BODY PLAN

Still a third group of worms differs from the roundworms in being *segmented*; that is, the body is formed as a series of short cylindrical units, connected together end to end and separated by partitions called *septa*. Many of the organ systems run continuously through the septa from one segment to the next. The earthworm or night crawler is a familiar and widespread example of these animals, which are classified as the phylum Annelida. An annelid worm is like a roundworm or nematode in having a tubular digestive system running the length of the body from mouth to anus, but the several organ systems are much more highly elaborated. In annelids, muscles surrounding the gut shield the gut from movements of the body wall. This is not true in roundworms, in which only the body cavity isolates the gut from the body wall. Circulatory and respiratory systems are present in a highly organized form in many annelids, together with relatively massive and specialized nervous systems. More elaborate excretory systems regulate the fluid composition of the body, and the digestive tract is considerably specialized from mouth to anus into separate chambers where particular digestive processes occur more efficiently than would be possible if they were dispersed and intermingled along the length of an unspecialized tube. (See Fig. 2-2.) Each segment includes a relatively large, fluid-filled cavity, in which the digestive tract and other organs are suspended, usually by membranes attaching to the organ itself and to the body wall. The advantages of segmentation seem to be largely associated with locomotion. In the case of the annelids, the connective tissue septa dividing the segments from each other are essential to the functioning of the "hydrostatic" skeleton discussed in the next section. In arthropods, the segments are often associated with appendages.

The circulatory system of annelids such as earthworms is almost as specialized in many ways as that of larger and more complex animals. Blood is pumped by peristaltic contractions of tubular blood vessels. The larger blood vessels branch to form thin-walled tubules in proximity to active cells to which the blood brings needed materials. Some annelids have appendages that extend a short distance outward from several of the body segments and are used to help the animals move, either by swimming or crawling. In certain species, these appendages are much longer and branch into thin tendrils filled with small blood vessels near the surface. These serve as gills, which take up oxygen from the surrounding water much more efficiently than could be achieved through the limited surface area of the body proper.

## THE ARTHROPOD BODY PLAN

A segmented body plan, containing specialized organs for all the major functions, has proved its worth in the great number of very abundant animals that have lived and multiplied with this arrangement. A very effective system of support has been attained, by the arthropods, by addition to the annelid body plan of a stiff *exoskeleton* secreted around the outside of each body segment. Primitive arthropods were aquatic and supported by the buoyancy of water. The exoskeleton, nevertheless, provides support for those forms that have become terrestrial. Perhaps more important, however, is the point that a rigid skeleton permits better muscle attachments and more rapid and more precise and discrete muscle movements. Rapid locomotion today is almost always associated with the possession of a firm skeleton, as is found in arthropods or in vertebrates and is conspicuously restricted in annelids. The arthropod skeleton, by its surface position, also provides protection. There is also an increase in length and complexity of the appendages with a jointed exoskeleton surrounding each segment of each leg.

Also important has been the "inverse waterproofing" of the body surface of many arthropods which prevents them from losing water from their bodies when surrounded by dry air. This feature has made it possible for them to join the vertebrates in being the major animal inhabitants of dry land environments.

The phylum Arthropoda, built along these lines, is by far the largest group of animals that has ever existed, in terms of number of species. The arthropods are certainly our most serious competitors for the resources of the earth. This phylum includes insects, spiders, and crustaceans, such as lobsters, crabs, and shrimp. (See Figs. 2-4 and 3-12.) The major organ systems are all present and well developed. Respiratory systems, in particular, show some unique specializations, which will be described in Chapter 5. The nervous system is a complex one, permitting a variety and complexity of behavior. The interaction of the nervous system with the endocrine glands is also exceptionally well developed in this group.

The phylum Arthropoda shows a variety of body forms which are passed through *en route* to the adult form; see *Animal Adaptation* in this series. Thus in beetles, butterflies, or houseflies, for example, the eggs hatch as wormlike *larvae*, caterpillars, or maggots, respectively, which are sexually immature, elongate, and differ from the adult in locomotion, feeding behavior (and even nutritional needs), sensory equipment, and often respiratory and regulatory mechanisms. Several *instars* or stages of larval life usually succeed one another, distinguished chiefly by increase in size but often with morphological changes and/or behavioral changes such as switches to different foods or moving from treetops to underground habitats. The last larval instar *metamorphoses* (literally changes its form) into a relatively quiescent *pupa*, often protected by a self-spun or constructed *cocoon*. The pupa undergoes

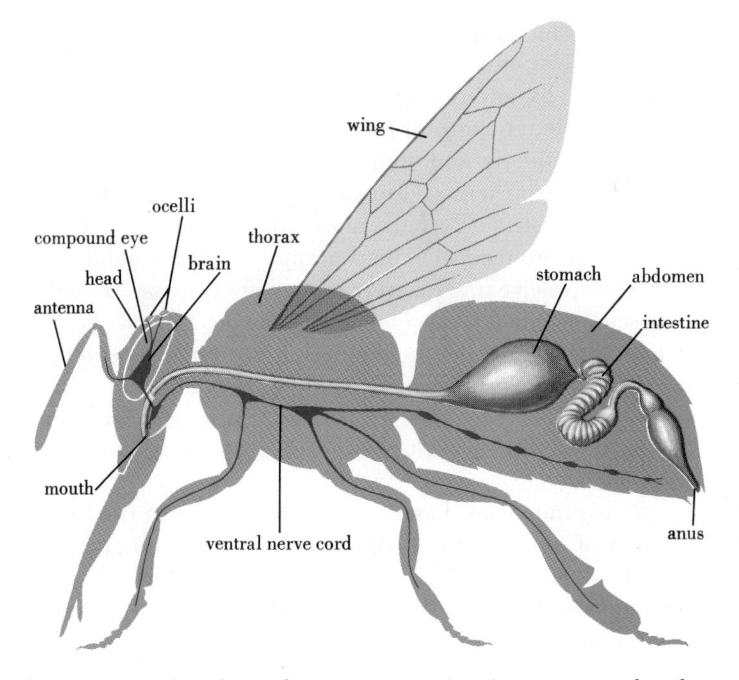


*Fig. 2-4 The body plan of a representative insect, a worker honeybee. The body cavity of insects is a haemocoel in which blood circulates to and around all of the organs. The complex anatomy of the haemocoel is not shown here. Note the appearance of limb and wings. (After Weber.)*

dramatic external and internal reorganization and metamorphoses after some days, weeks, or months into a sexually mature, flying *adult*. Such changes in body plan and physiology are strategic solutions to life's challenges. Examples of such surprisingly different life stages are known from almost all other phyla as well.

## THE MOLLUSCAN BODY PLAN

The most familiar mollusks are the rather sedentary clams and oysters, whose body plans seem to be a distinct comedown after even the briefest consideration of the structure and behavior of such arthropods as the insects. There are, however, certain members of the phylum Mollusca, such as snails and squids, that are surprisingly complex. The snails are notoriously slow travelers but their internal organs are highly specialized; the class Cephalopoda, which contains the squids (and the octopuses), represents an extreme specialization of the molluscan body plan for an active, free-swimming life. The cephalopods are compactly formed, with a solid body containing a digestive tube with mouth and anus, a reproductive system, a chemical-regulating excretory system, a nervous system, and in most cases respiratory organs. In mollusks in general, there are no jointed appendages comparable with those of arthropods, for instance; instead there is a solid mass of muscle, called the foot, that serves for locomotion by a complicated series of partial contractions of its various sections. Its actions resemble that of the human tongue in the variety of effective shapes it can assume. In the squids and octopuses, the foot becomes subdivided into separate arms that move freely by well-controlled muscular contractions. These in turn require precise regulation, which is provided by a highly developed nervous system. The eyes of an octopus seem, in structure and function, to rival those of mammals; the brain of an octopus is larger and more complicated in structure and function than that of any other group except the vertebrates. Basically, mollusks are constructed on a plan that entails compactness, the absence of paired appendages and obvious segmentation, and, finally, the secretion of a hard shell. The latter is prominent in clams, snails, and similar mollusks; in the squid and octopus, the shell is largely internal and not nearly so conspicuous. (See Fig. 2-5.) Paleontological evidence indicates that mollusks are derived from segmented ancestors. A recently discovered mollusk, *Neopilina*, is bilaterally symmetrical and segmented. Its body plan somewhat resembles that of annelids.

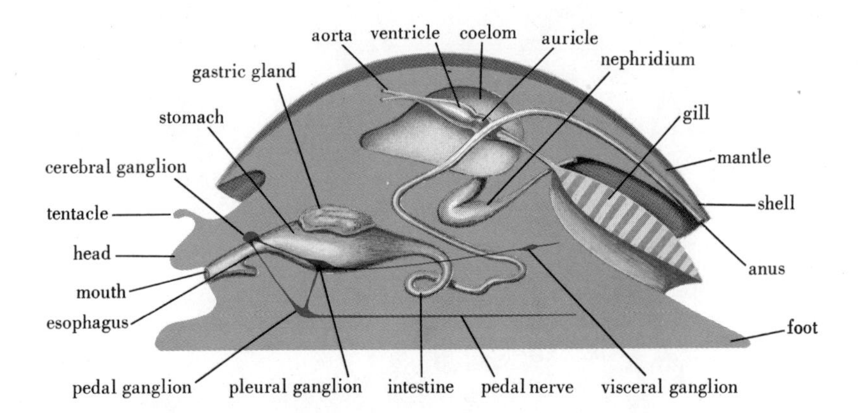


***Fig. 2-5*** *The body plan of a hypothetical ancestral mollusc. A portion of the coelom is shown which surrounds the heart and from which the excretory nephridium takes its origin. The ctenidium is a typical molluscan gill for respiratory gas exchange. (After P. Meglitsch,* Invertebrate Zoology, *Oxford University Press, 1967.)*

## THE ECHINODERM BODY PLAN

The invertebrate phyla can be divided, in general, into two great evolutionary groups. Those phyla which have already been described tend to fall into one group; a second group, from which vertebrates took their origin, also gave rise to the echinoderms. One of the most striking features relating the echinoderms to the ancestors of the vertebrates is the similarity in body form of the larva which is often, among the echinoderms, the only really mobile life stage. Echinoderm larvae do not move actively over large distances but are carried passively by currents and distributed far and wide. Most living species of echinoderms are characterized as adults by radial symmetry with well-defined *oral* and *aboral surfaces*, as in coelenterates. The larval forms, as well as some extinct echinoderms, exhibit bilateral symmetry. Most

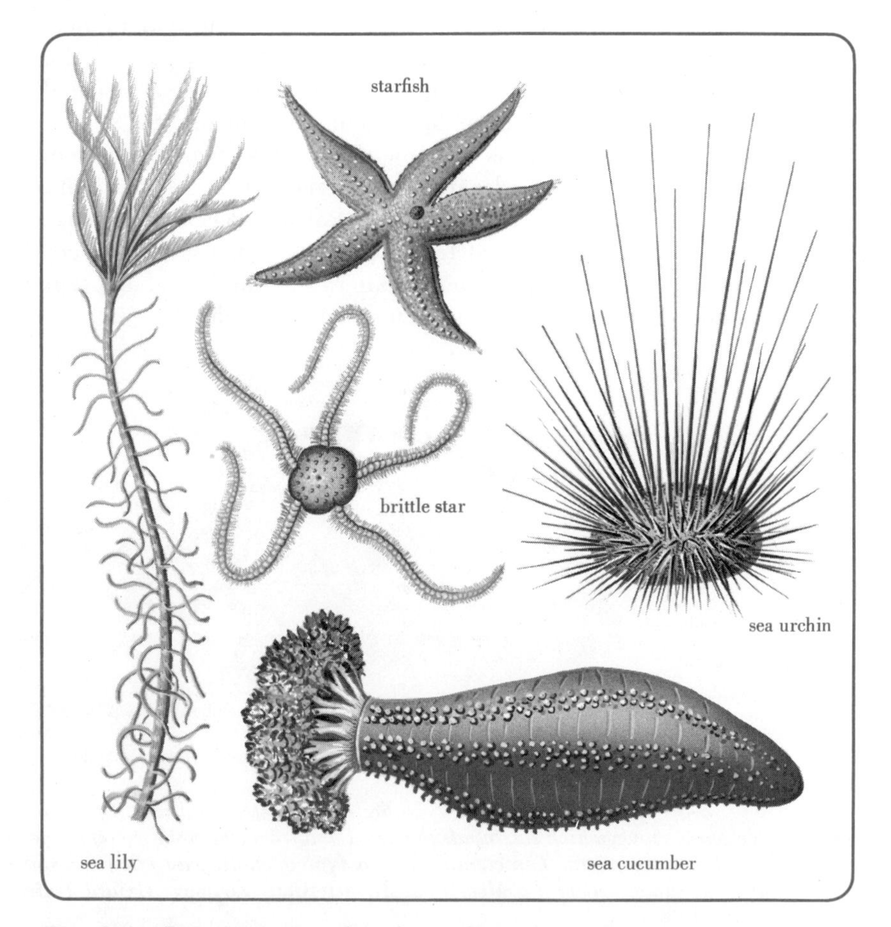


***Fig. 2-6*** *Representative echinoderms. Note the predominantly pentamerous (five-armed) radial symmetry.*

adult, living echinoderms have five arms, or arms which occur in multiples of five, surrounding a disklike central portion that includes the mouth and anus. (See Fig. 2-6.) The echinoderms have a unique internal water-filled body cavity, referred to as the *water-vascular system.* It serves both as a circulatory system and as the power supply of the locomotor system. Although the adults of many of the species in this group are known for their relative quiescence (many are sessile), other echinoderms such as the starfish move by adjusting the hydrostatic pressure in hundreds or thousands of *tube feet.* Primitively these tube feet were used as tentacles for bringing food to the mouth, but they have often been specialized in more advanced forms for locomotion. The digestive and regulatory systems are often well developed in this group. The respiratory system is usually not particularly specialized.

## THE VERTEBRATE BODY PLAN

The remaining major pattern into which the organ systems of animals are arranged is similar to our own and that of a host of animals we recognize as more closely akin to us than insects, squids, or worms. Its most obvious feature is a jointed skeleton, deeply internal, in contrast to the external, jointed skeletons of the arthropods and the one- or two-piece skeletons of the mollusks. We share this body plan with seven clearly distinguishable groups of animals, all more or less familiar. These are (1) the other *mammals*, which also nurse their young, bear their young alive (with a few interesting exceptions), have hair or fur as a body covering, and regulate their body temperature; (2) the *birds*, which hatch their young from eggs, are covered with feathers, regulate body temperature, and are usually highly specialized for flight; (3) the *reptiles*, which have scaly skins, exhibit limited (principally behavioral) regulation of body temperature, have teeth that are not so specialized from one part of the mouth to another as in mammals, and lay birdlike eggs protected by a tough shell and specialized in such a way as to allow the embryos to develop on land; (4) the *amphibians* (frogs and salamanders, for example), which have a moist skin and must lay their eggs in water or in very high humidity because they lack embryonic membranes that can withstand drying. Most members of these four groups, known as *tetrapods*, have the familiar two pairs of limbs, usually for locomotion, and all have basically similar organ systems, normally including lungs for breathing air. Some of the amphibians also have gills somewhat like those of fish. The fifth, sixth, and seventh groups are fishes, which lack the walking limbs of tetrapods and normally have elaborate gills in place of lungs in order to obtain oxygen dissolved in water. Among the fish, there are three living groups that are sufficiently different that they are placed in separate classes. The most primitive are the lampreys and hagfish which lack jaws and paired fins and exhibit strikingly primitive skeletons, sense organs, and kidney development, among other things. The second group, the sharks and rays and their relatives—the *cartilaginous fish*—differ sharply from the third group, the *bony fish*, by their lack of bone and rather different organization of almost every organ system but share with the bony fish the very important development of jaws and paired fins. The amphibians are a transitional group intermediate between almost purely aquatic fishes on the one hand and terrestrial reptiles on the other. They are believed to have arisen from bony fish only remotely related to currently living forms.

Almost all of these vertebrate animals, except lampreys, hagfish, and cartilaginous fish, have the same type of internal skeleton composed of *bone*, which is itself a unique vertebrate material, formed by the secretions of thousands of specialized cells. Although hard and rigid, bone is more clearly a living organ than the secreted skeleton of mollusks. In the exceptional groups that do not have bone, *cartilage* performs many of the same functions as bone and in much the same fashion. Extinct relatives of even these cartilaginous fish, known only as fossils, had, however, already evolved the typical vertebrate bone structure. In the two living boneless groups apparently bone has been evolutionarily replaced by cartilage. The bony skeletons of vertebrate animals are jointed and the several bones are connected by appropriate types of living material to give the whole structure a flexibility that allows for a variety of movements and behavior.

The mutual relationships between the major organ systems in vertebrates are also quite different from those found in the arthropods or in other phyla. The nervous system not only is the most complicated found in any group of animals, but also is larger and more specialized. It is located dorsal to the digestive tract rather than ventral, as in the arthropods, for example. A dorsally located nervous system is not known to be "better" but, like the flatness of the flatworms, it provides a convenient distinguishing feature. The four limbs of the terrestrial vertebrates are also supported by the same sort of jointed, internal, body skeleton as the rest of the body, and although these limbs have become highly modified in many of the different kinds of vertebrate animals (or even lost altogether as in the snakes), they are built according to a similar plan, as illustrated in Fig. 2-7.

Vertebrate animals are all members of the phylum Chordata, along with a few other groups of much simpler animals that share with some of the more primitive fishes and with embryos of other vertebrates a type of axial skeleton called the *notochord*. This long, cylindrical rod, surrounded by connective tissue and consisting of specialized cells found only in the notochord, is located dorsal to the digestive tract

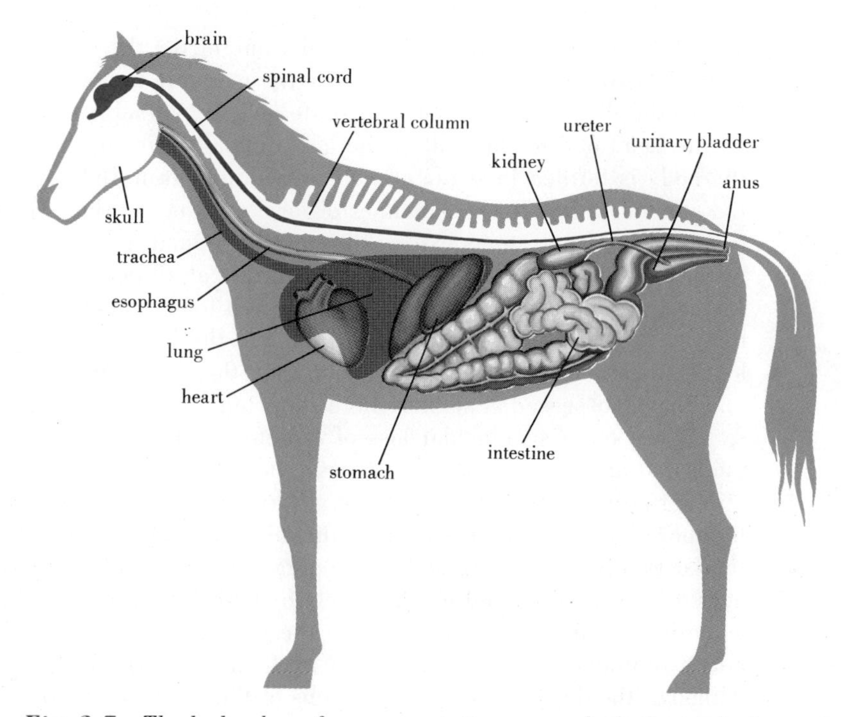


***Fig. 2-7*** *The body plan of a representative mammal, the horse. An internal skeleton (only partly shown) supports the body and encloses and protects the central nervous system. The digestive and respiratory systems are enclosed in and separated from the body wall by the coelom (not shown).*

but ventral to the spinal cord. The notochord is believed to prevent the animal from contracting in length (as an earthworm does) during locomotion. The typically elongate chordate body is designed, rather, for sinuous or side-to-side movement. Vertebrates are also characterized by bilateral symmetry; cephalization; the presence of a dorsal, hollow nervous system; the division of the body into head, pharynx, trunk, and tail; and the presence, at some stage, of internal gills.

## DEPARTURES FROM THE MAJOR BODY PLANS

This chapter has merely sketched, in broadest outlines, the body plans of the most successful and numerous phyla of animals. Many other phyla have not been mentioned, and many members of the phyla discussed have become so modified in the course of their evolutionary history that the characteristic body plan is obscured. These modifications are particularly evident in animals that have become specialized for a parasitic way of life or for life as attached or sessile forms.

To cite only one extreme example, one of the three subdivisions of the flatworms (the class Cestoda) is the parasitic tapeworms. These live in the digestive tracts of vertebrate animals, attaching themselves by the head end to the wall of the host animal's intestine. The rest of the body is formed in segments, but these differ from the segments of the annelid worms in their manner of formation and in that no circulatory or digestive systems extend through the animal from one segment to the next. Indeed, there is no digestive tract at all in the tapeworms. Living in an environment filled with digested food molecules, they obtain all the food they need directly through the body wall; they have lost, altogether, the digestive cavity that their presumed ancestors must have possessed at some remote stage in their evolutionary history.

Even more spectacular loss of structure is found in a few of the mollusks and crustaceans that have become parasitic inhabitants of other animals. Although their larval forms identify their relationship to other mollusks and crustaceans, the adult form that develops within the host animal is little more than an organ that produces the parasite's eggs or spermatozoa and is completely nourished and protected by the activities of the host. In some cases the parasite even inactivates the host's reproductive organs, perhaps to "conserve" the host's energy. Although the degenerate specializations of these parasites are extreme examples, and other parasitic forms (among the nematodes, for example) are structurally similar to their free-living relatives, in studying particular animals one must be prepared to find the basic body plans that have been described in this chapter modified and altered to suit thousands of special needs.

## *DIVERSITY OF ENVIRONMENTS*

The environments inhabited by animals may, at first, be classified into the categories saltwater, freshwater, and terrestrial. These are, however, by no means uniform in their characteristics or in their challenges to life. The oceans are, in many ways, the most stable of the available habitats. The degree of salinity varies from place to place depending upon evaporation rates and the inflow from great river basins, but in a given portion of the ocean, the salinity is rather stable as are the temperature, the clarity of the water (and, therefore, the depth to which light reaches and the depth to which photosynthesis is possible), the oxygen concentration, and the supply of crucial nutrients, such as phosphates. The productivity of oceanic waters in terms of *biomass* (the total weight of living organisms per unit volume) is very different from place to place. The colder waters (in which $O_2$ is more soluble) of the North and South Atlantic, for example, are sufficiently turbulent to stir up nutrients and to mix dissolved $O_2$ to great depths and are so productive of life (from planktonic invertebrates to fish and whales) as to be semiopaque during some seasons. Great industries are based on their productivity. Some tropical parts of the ocean are relatively sterile, apparently in part because of the lower solubility of $O_2$ in warmer waters and the lack of turbulence.

In such marine environments, animals exhibit a variety of interesting and essential adaptations. Their body fluids, in general, are less saline (or have a lower osmotic pressure) than salt water. They tend, therefore, to lose water to the ocean through their surfaces and become dehydrated. Freshwater organisms have the opposite challenge; they tend to become waterlogged. Some of the solutions to these problems are discussed in Chapters 3 and 8.

Food is normally abundant in the oceans, but so are predators, and visibility or exposure is often rather high. Among species living in open water, patterns of coloration that decrease visibility are common. Often fish in such waters are dark above and pale below so that seen from above against the depths or seen from below against the bright sky they are obscured. Many fish are transparent or translucent; others are silvered so as to reflect the light. Such reflections can be very confusing to the eye, especially when species that aggregate into schools are involved. Schooling, a common behavior pattern among oceanic fish, seems to function to confuse predators, which find it difficult to pursue an individual fish persistently as their attention is distracted by hundreds of alternate targets.

Many invertebrates—and fish, as well—engage in daily vertical migrations, downward at dawn and upward at dusk. This practice may well reduce their vulnerability to predators that hunt by vision or may be for energy conservation. Many other species, on the other hand, are adapted for living below the lighted zone, in perpetual darkness. Here their vulnerability to predation may be reduced, but they are largely dependent for food on the productivity of the lighted zone above. Some are scavengers on the rain of dead organisms from above, sometimes living in the decaying ooze of the ocean floor. Yet others prey upon their deep-sea colleagues but, since meals are few and far between, are adapted (by cavernous mouths and digestive tracts) for taking massive meals when the opportunity presents. Living at such depths poses several additional special problems. First, it may be difficult to find and recognize a mate in the dark. Species and sex recognition is often provided for, in fact, by patterns of luminescence. Even more startling, in one group of fish, the male is actually parasitically attached to the female from early life so as to be available for reproduction at all times. Equally challenging has been the problem of adaptation to the high pressures of the deep seas. A variety of buoyancy-altering adaptations have been recognized but, undoubtedly, many special solutions remain obscure.

Many oceanic species have specialized for life along reefs or on the continental shelf where cover is available. Here one finds an abundance of brightly colored, conspicuous species and also a grand array of fantastically camouflaged forms. Indeed, one finds close community associations often among different phyla. An interesting relationship, not entirely "voluntary" on one side, exists between certain crabs and sea anemones. These crabs select anemones of appropriate species and suitable size and, clasping them with one pincer, carry them around. The crab presumably gains camouflage value. In other cases, small, brightly colored fish shelter among the tentacles of sea anemones or other coelenterates, where they apparently gain protection from the proximity of the coelenterate's nematocysts and where the coelenterate may share food captured by the fish.

The intertidal zone of the shores and the mud flats of shores and estuaries make up yet another complex array of habitats. Oxygen may be scarce or absent in the mud because of the high rate of decay. Water cover may come and go with the tides. Temperature fluctuates widely. Turbulence and mechanical abrasion may be strong. Vast numbers of species of animals of many phyla are adapted for life in such zones. The abundant food supply may make up for many of the difficulties.

Freshwater habitats are neither uniform from place to place nor from time to time. In cold, turbulent mountain streams, $O_2$ is abundant. The swift current and the sudden effects of storms must, however, be coped with. Food is also sparse. In tropical swamps, $O_2$ may be totally absent below a millimeter or so of the air-water interface. Not only is $O_2$ poorly soluble in hot water, but there is little stirring of the water and photosynthesis is commonly reduced by the opacity of the water produced by organic products of decay coming from a bottom layer of decaying debris (in which any $O_2$ is rapidly consumed). The adaptations required for life in such hot, $O_2$-poor, quiescent swamps are clearly different from those required in swift-flowing, cold, $O_2$-enriched streams. Consider, for example, the challenges provided for one part of the fish life cycle, the egg. In mountain streams, the eggs must be anchored if they are to stay in the same habitat. In some cases they are buried in the sandy bottom or deposited on the underside of rocks. Hatching is likely to be slow because of the low temperature. The young often have to be predators, since vegetation can be quite sparse and microorganisms are swept away; thus larger, yolkier eggs may prevail.

In tropical swamps, on the other hand, the eggs may be floated in bubble nests on the surface, where they have atmospheric $O_2$ available. Or they may be fixed to green, floating, photosynthesizing leaves which also assure a local $O_2$ supply. In other cases, the eggs appear to be actively aerated by a reverse gill action of the parents. The parents breathe atmospheric air (see Chapter 5) and give off $O_2$ to the eggs. The eggs hatch rapidly and the often very small young in such swamp species feed on microorganisms and vegetation.

In other freshwater environments, fluctuations with weather can be extreme—temperature, water flow, and exposure to sunlight, for example. Yet other freshwater habitats, such as deep caves, may be exceptionally stable.

Finally, there are terrestrial habitats. Even here there are niches that are dark, damp, and sheltered, such as under rocks or in the leaf mold covering the floors of forests in the temperate zones. There are the exposed, dehydrating niches of the open desert or mountain tops or aloft in the air, and there are the secretive, protected niches of the jungle canopy or at ground level in the prairie. The array and spectrum of challenges is vast. Often, but not necessarily, fluctuations of temperature, humidity, wind, and sun exposure are relatively large. The types of food that are available and the opportunity to exploit these supplies are also very variable. It is practical, for example, to be *frugivorous* (feeding on fruit) in the tropical rain forest, where some varieties of fruit are ripening year round. A diet of seed is practical in the prairie, and the pursuit of insects in the night air is practical during summers in temperature climates. But feeding on fruit in New England (by foxes, bears, deer, and robins, for example) has to be limited to one or two months at best. Seeds (except those held aloft) offer little sustenance to birds (though they may to burrowing rodents) when the ground is deeply covered by snow. The insectivorous bats of temperate climates must migrate or hibernate in winter, since few insects are then abroad.

We have space here only to introduce these ideas about the variety of environments and how they place demands on the evolutionary ingenuity of animals. Amazing niches such as hot springs, the ocean abysses, deserts, and the artic tundra have, indeed, come to be inhabited, proving the potential that animals have for structural and functional adaptation. We shall deal with some of these specializations in later chapters. We shall also, however, emphasize the generalities of animal form and function where these have been recognized. The interplay of these two themes—general biological solutions to life's problems and special solutions to special challenges—make up much of biology. Often the generalities can only be recognized after many specialized and often esoteric-seeming special cases are analyzed. Rules are assembled out of such examples. Themes are traced through arrays of specific cases. The ultimate goals of such studies, of course, vary from scholar to scholar. The goal may be to unveil a bit more of truth; to delineate, describe, and analyze more systems and interactions. The goal may be better to understand man, his nature, obligations, and destiny. In either case, the pathways are much the same. We have to study as wide a variety of organisms as possible, the problems they face, and the solutions they have found. But we must also explore, in depth, the principles of such problems and solutions, and we do so by choosing as experimental subjects species that are amenable to our purposes and approximate generality in some of the solutions they display. The honor roll of such species includes the rat, the guinea pig, the frog, the fruitfly, and the bacterium *E. coli*. From such compendia of descriptive and experimental knowledge we hope better to grasp the nature of our own existence.

## FURTHER READING

Barrington, E. J. W., *Invertebrate Structure and Function*. London: Nelson, 1967.

Beament, J. W. L., and J. E. Treherne (eds.), *Insects and Physiology*. New York: Elsevier, 1968.

Clements, A. N., *The Physiology of Mosquitoes*. New York: Pergamon, 1963.

Croll, N. A., *Ecology of Parasites*. Cambridge, Mass.: Harvard University Press, 1966.

Dales, R. D., *Annelids*. London: Hutchinson University Library, 1963.

Etkin, W., "How a Tadpole Becomes a Frog," *Scientific American*, 214 (5): 76–88, 1966.

Jensen, D., "The Hagfish," *Scientific American*, 214 (2): 82–90, 1966.

Lanham, U., *The Insects*. New York: Columbia University Press, 1964.

Loewy, A. G., and P. Siekevitz, *Cell Structure and Function*, 2d ed. New York: Holt, Rinehart and Winston, 1969.

Marshall, A. J. (ed.), *Biology and Comparative Physiology of Birds*. Vols. 1 and 2. New York: Academic Press, 1961.

Mayr, E., *Animal Species and Evolution*. Cambridge, Mass.: Harvard University Press, 1963.

Meglitsch, P. A., *Invertebrate Zoology*. New York: Oxford, 1967.

Morton, J. E., *Molluscs*, 3d ed. London: Hutchinson University Library, 1964.

Romer, A. S., *The Vertebrate Body*, 3d ed. Philadelphia: Saunders, 1962.

———, *Vertebrate Paleontology*. Chicago: University of Chicago Press, 1966.

Rothschild, M., "Fleas," *Scientific American*, 213 (6): 44–53, 1965.

———, and T. Clay, *Fleas, Flukes, and Cuckoos*. London: Collins, 1952.

Runcorn, S. K., "Corals as Paleontological Clocks," *Scientific American*, 215 (4): 26–33, 1966.

Satir, P., "Cilia," *Scientific American*, 204 (2): 108–116, 1961.

Sleigh, M. A., *The Biology of Cilia and Flagella*. New York: Pergamon, 1962.

Schmitt, W. L., *Crustaceans*. Ann Arbor: University of Michigan Press, 1965.

Stanier, R. Y., M. Doudoroff, and E. A. Adelberg, *The Microbial World*, 2d ed. Englewood Cliffs, N.J.: Prentice-Hall, 1963.

Thimann, K. V., *The Life of Bacteria*, 2d ed. New York: Macmillan, 1963.

# *chapter* 3

# *Structural Systems*

Protoplasm is more than two-thirds water, but living organisms are far from being liquid. Not only do they maintain a fairly definite shape, but each one keeps in its body fluids a chemical composition that differs, often markedly, from the surrounding medium in which it lives. The structural systems considered in this chapter are those parts that enable animals to hold a particular shape, both when at rest and when engaged in their various activities. Structural systems may also provide mechanical (and chemical) protection. In addition, skeletal elements transform muscular activity into effective movements. It is convenient to begin with aquatic animals, not only because the earliest living organisms almost certainly lived in sea water, but because their problems of support are the simplest.

### *STRUCTURAL INTEGRITY IN AQUATIC ANIMALS*

Any animal living in the water has an immediate advantage over one that lives in air, in that protoplasm has almost the same density as water and, hence, the structure and shape of an aquatic animal are not constrained by limitations imposed by gravity. A jellyfish, for example, having much the consistency of a very thin, gelatin dessert, will, when stranded on a beach, flow and droop into a flat blob bearing little resemblance to the beautiful shape that floats in the water. A large whale, although it is a mammal with a sturdy internal skeleton, is unable to hold its shape adequately if stranded by the falling tide and usually suffocates because its supporting system is poorly suited for holding in place several tons of soft flesh without the aid of surrounding water of nearly equal density.

Even though aquatic animals have relatively little need for support against gravity, they must all protect themselves from too rapid and free an exchange of their constituent molecules with the water around them. Since the same solvent, water, is present inside and out, a sodium ion or a glucose molecule near the animal's surface would diffuse either outward into the environment or inward depending upon the relative concentrations. Hence even the simplest aquatic animals possess some sort of skin that is relatively impermeable to water, to many ions, and to the small molecules that are distributed in the body. But an animal cannot afford to wall itself off behind a totally impermeable membrane, for it must obtain some molecules such as food and oxygen from the outside, and other substances must be eliminated lest they accumulate in excessive concentrations within the cytoplasm. The essence of a living surface is thus a balance between too much permeability and too little permeability, between allowing the escape of too much useful material and impeding too greatly the intake of needed molecules.

### *SKIN*

Multicellular animals universally have an outer covering of cells, normally derived from a special layer of embryonic cells called *ectoderm* and often combined with elements of another embryonic layer, *mesoderm*. The *skin* is as much an organ as the heart or the kidneys. In higher phyla, the skin is likely to serve many complex functions, but in all animals it serves basically to delineate and to isolate partially the body from the environment. All living organisms have intracellular fluids and almost all have extracellular fluids differing in constitution from the medium in which they live. The fluids of many organisms differ in osmotic pressure (a measure of the concentration of dissolved ions and nonionic molecules in the solvent, water) from the surrounding medium, whether sea water, freshwater, or any of a wide variety of variants of these. All organisms maintain concentrations, for example, of sodium, potassium, magnesium, chloride, and carbonate ions, glucose (or other sugars), amino acids, and oxygen, that are different from the concentrations found in sea or fresh waters. The cells themselves must be permeable to all of these and to many more ions and molecules in order to nourish themselves, maintain their internal environment, and rid themselves of metabolic waste products. An outer covering, therefore, is required by multicellular animals to isolate their contents from those of the environment, lest solutes diffuse freely in or out, across their respective concentration gradients.

Skin can be made waterproof and impermeable to ions and small molecules but only at the expense of isolating the animal completely. Some particles, such as bicarbonate and ammonium ions, may have to be released and others, such as glucose, oxygen, and sodium ions, may be needed. As a result, the design of skin has been a compromise between isolation and selective permeability. The greater part of the surface of many animals is, indeed, more or less impermeable to water, ions, and small molecules. Such impermeability has been achieved in various ways. Insects, for example, lay down extracellular waxy materials on the outer surface of the skin. (See Fig. 3-1.)

The price of a generally or relatively impermeable surface is the need for specialized regions for the absorption of nutrients, for respiration, for the adjustment of concentrations of necessary ions, and for the

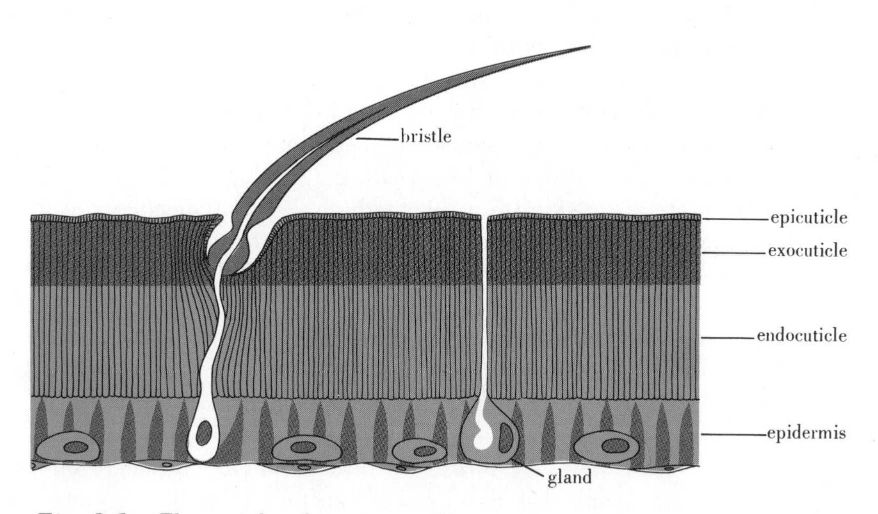


***Fig. 3-1*** *The cuticle of an insect. The cuticle contains chitin that stiffens the exoskeleton. (After Wigglesworth.)*

extrusion of nitrogenous wastes. Such permeable surfaces may be derived from the skin or from the gut, or jointly from both, and are often internalized for protection, closer regulation, and conservation of water. Thus one finds, among various vertebrates, oxygen and carbon dioxide exchange via external and/or internal gills, lungs, special branchial (pharyngeal) chambers, the lining of the stomach, intestine, or cloaca, the buccal mucosa, and the skin. In each such animal, a large part of the rest of the body's surface is impermeable and the same or other local permeable regions are specialized for the passage of other solutes or water. An even greater variety of specializations for localized permeability is found among the invertebrates.

Thus one can think of an animal as an isolated body with an impermeable surface, often with a deep invagination or tubelike penetration (gut) of the environment. Some portions of the external and/or invaginated surfaces must be permeable.

Skin also serves to protect the body mechanically. The skin may be reinforced with bone (as in the scales of the bony fishes), or chitin (as in insects), or covered with mucus (as in hagfish), or fur or feathers or leathery scales, or in many other ways may be designed to resist abrasion or penetration. The stiffening or reinforcing mechanisms may also give shape to the body or may even provide attachments for muscles, as in the arthropods.

Skin also helps to exclude bacteria and parasites from the body, but many such organisms have circumvented this problem of entry by taking advantage of the penetrating bites of mosquitoes or tsetse flies, for example, to enter the bloodstream or by entering via vulnerable permeable areas. Some parasites even make a direct attack on the skin and abrade their way into the body.

In those animals that regulate body temperature (birds and mammals, for example), skin may serve to insulate the body by its outgrowth of fur or feathers or to conserve or dissipate heat by changes in blood flow near the surface or by sweating. (See Fig. 3-2.) Fur and feathers also protect the skin from solar radiation and other effects of weather. In many mammals and birds, the degree of insulation may be varied by fluffing of feathers, erection of the hairs of the fur (represented by "goose pimples" in man), or by growing a heavier or lighter coat in keeping with the season. Even here, the need to balance isolation with permeability can be seen where special areas of skin or the surface of the respiratory tract may be used for heat dissipation by evaporation of water. In some reptiles, specialized vascular skin may be used to absorb solar heat. In birds, the brood patch, a specialized area of vascularized, naked skin, is used to transfer heat to the eggs or nestlings. The total surface area of skin may be increased either for the absorption or dissipation of heat, as in the giant ears of elephants.

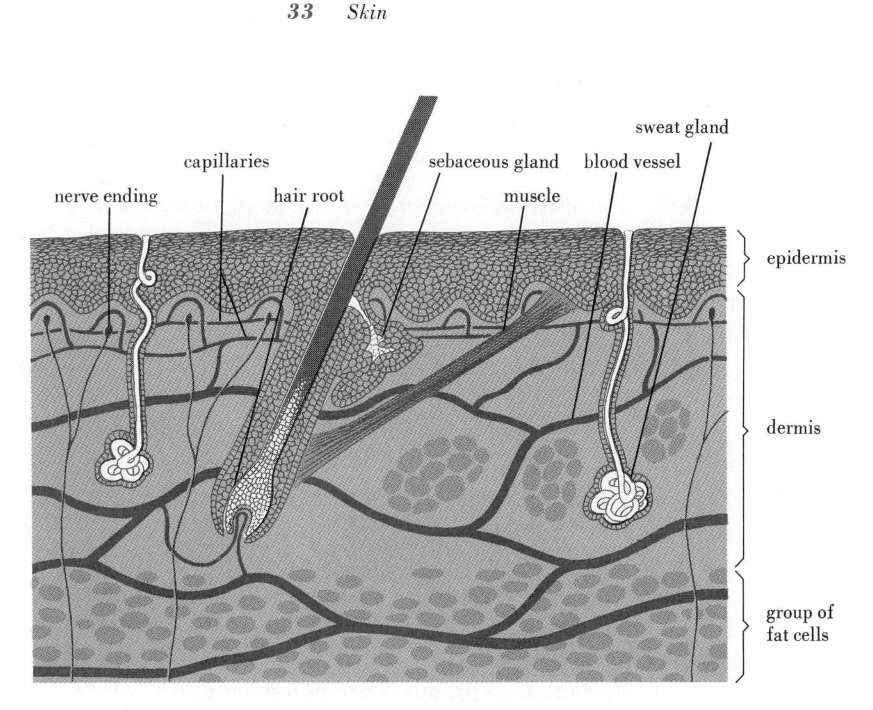


***Fig. 3-2*** *The skin of man.*

Pigments have also been exploited for greater efficiency of heat absorption or radiation. Pigmentation of the skin has also been used widely for camouflage, warning, mimicry, and species or sex recognition, and for protection against ultraviolet light. Characteristic markings are exploited, for example, by many bony fish for indicating sexual maturity and reproductive readiness and to release courtship behavior in potential mates. The degree of pigmentation may also vary on a day-night cycle—as, for example, in crabs, which darken at night and lighten by day. Such a cyclical variation with a period of about one day is called a *circadian* (from the Latin, meaning "about one day") *rhythm.* Such features of skin may be in response, presumably, to the evolutionary pressures of light and vision in predators or parasites. Skin has also responded to other senses (in predators or potential mates) and other physical phenomena. Chemical products of dermal glands may be used for example to mark territories (common in mammals), or for species, colony, or sex recognition.

Skin may also be the location, as one might expect, of many sense organs—those for taste and other chemical senses, neuromasts (sense organs that occur in the inner ear of vertebrates and the lateral line system of fish; see Chapter 10 and Fig. 10-22), and those for light reception. Even the complex vertebrate eye is partly derived from skin.

Clearly the surface of an animal is not only a convenient site for sensation but for the acquisition of food. Many skin derivatives have evolved as parts of feeding mechanisms. Obvious examples are vertebrate teeth and many of the different straining and trapping devices of crustaceans.

Skin has been used even more conspicuously and subtly for communication. In primates, facial expression, achieved by fine muscles (derived from the ancestral gill musculature) attached to facial skin, conveys dozens of emotional states and functions closely with vocalizations or speech in communication. Other examples of communication via skin would be changes of color in sexual skin in monkeys with the estrous cycle, blushing in man, or the change in shape and silhouette in dogs or cats when the fur is erected in anger or in fear.

Skin may even be designed for relative nonconductivity of electricity, as in some electric fish where specialized conducting pores apparently lead through the otherwise insulating skin to the electrosensitive sense organs. (See Chapter 6.)

Finally, opacity versus transparency should be noted. Transparency may improve camouflage in aquatic animals. Opacity protects against the physical effects of light and ultraviolet light, especially on the nervous system. Compromises are common in fish, where, although the animal as a whole is relatively transparent, the central nervous system is often covered by reflective silvery tissue or by black pigment.

This by no means completes the catalog of functions that have evolved for skin in the animal kingdom. Again because of its being the outer covering, skin has often been specialized for fighting and aggressive defense. The development of weapons, whether they be as small as the nematocysts of coelenterates or as large as the horn of the rhinoceros, has been quite common. (*Nematocysts* are intracellular weapons, usually coiled, threadlike organelles, peculiar to coelenterates. Contained in specialized surface cells called cnidoblasts, they are discharged when the surface is brushed by another organism and often either entangle small organisms or penetrate larger ones and inject a disabling poison into them.)

## SKELETONS

Skin alone is not adequate support for most animals. Internal and external skeletons are found even in some of the protozoans and, among the coelenterates, the corals are famous for their externally secreted skeletons. None of the worms—flat, round, or segmented—have developed important skeletons, although some of the sedentary annelid worms secrete material around their bodies in the form of protective tubes. The annelid worms and the echinoderms have specialized and quite different "skeletal" systems, which are partly hydraulic, depending on the encapsulation of water or body fluids. The three most advanced phyla—the Mollusca, the Arthropoda, and the Chordata (principally the subphylum Vertebrata)—have complex skeletons, each employing a radically different plan. The differences among the three may conveniently be analyzed in terms of the material from which the skeleton is formed, the anatomical position of the skeleton relative to the other organs of the body, and the arrangements for articulation of the separate elements of the skeleton with one another.

The skeletons of mollusks are mainly composed of calcium carbonate ($CaCO_3$), a ubiquitous substance that takes on quite different properties in the shell of a clam or snail from those of $CaCO_3$ in marble or limestone. The molluskan shell is secreted progressively, in layers, by cells on the outer surface of the body as the animal grows. Protein fibers are also laid down between the layers of $CaCO_3$, and other organic substances are interspersed in smaller concentrations to give the whole structure a considerably greater strength per unit weight than would be found in pure $CaCO_3$. The shells usually have an outer layer of more horny material, which contains a greater proportion of protein. Pigments may be present to give the shells a variety of patterns and colors. These are especially apparent in snails.

Although to a chemist a clamshell may seem to be little more than $CaCO_3$ (with a few traces of "impurities" such as proteins), a study of its microscopic structure reveals an elaborate organization reflecting its secretion by living cells. The organic "impurities," indeed, are the key features. The biological organization of the shell substantially increases its usefulness for its purpose of supporting and protecting the body that formed it. The shells of mollusks are enlarged by the addition of new, secreted material near their outer edges. The cells lying just under the outer lip of a clamshell lay down new material—rapidly when food is plentiful and conditions favorable, more slowly if times are hard. Other cells lying just under the older part of the shell, nearer to the hinge, add more material to increase the shell's thickness; the patterns in which $CaCO_3$ and protein fibers are laid down differ according to the age and thickness of the shell.

Arthropods form their external skeletons as plates on the outer surface of the body proper and its appendages. These plates are composed of a hard material secreted by a layer of living cells just beneath the exoskeleton. The exoskeleton of an insect may be used as an example, as other arthropods differ only in relatively minor ways. As illustrated in Fig. 3-1, the living cells that form the hard covering are a single layer, most of them relatively unspecialized. Each one secretes a variety of materials as the *cuticle* is formed. Last to be secreted, but outermost, on the finished surface, is a waxy layer with protein interspersed; underneath this, and formed somewhat earlier, is a harder layer containing not only the ubiquitous fibrous proteins but a material called *chitin* that is especially characteristic of insect exoskeletons. The waxy layer is secreted through pores in the integument. Although found in the hard parts of several phyla of animals, only in the arthropods is chitin widely used to strengthen a jointed exoskeleton. Chitin is chemically related to sugars and polysaccharides, being composed of many units having the same molecular structure as glucose but with an $NHC{=}OCH_3$ or acetylamine group replacing one of the OH groups of the glucose molecule. Chitin is insoluble in every ordinary solvent, as is cellulose, the major material of plant cell walls—the material that is processed to make cotton cloth, paper, and a variety of other useful materials. Cellulose is made up of a polymer of glucose units in the same way that chitin is a polymer of similar groups containing nitrogen. Both cellulose and chitin are very resistant to digestion or other biological deterioration and persist for a long time after the death of the organism. Nevertheless, chitin is digested by insects themselves at each molt.

In the exoskeletons of insects and other arthropods, the chitin is closely bound to protein molecules. This combination gives the uniquely strong, flexible, and resilient properties of the actual skeleton of the living animal. The chitinous plates are contoured to form an efficient armor—thinner where bending occurs, thicker and reinforced by spines and ridges where stiffness is required. Furthermore, at intervals, there are small holes through which protrude horny, stiff hairs; these are not rigidly fused to the exoskeleton itself but are free to move slightly in a socket formed by one or more of the specialized cells surrounding the cell that forms the hair. Often there are nerve cells closely associated with hairs; these may be sensitive to minute movements of the hair, sometimes even to the submicroscopic deflections caused by sound waves or by vibrations of the ground or vegetation signaling the approach of some other animal, or they may serve the senses of taste or smell. At the joints between the plates of an insect's exoskeleton, the hardened *exocuticle* is absent, but the thinner, more flexible *endocuticle* and the waxy, leatherlike outer layers of the cuticle are continuous through the region of the joint. Thus an insect's skin is a continuous one over all parts of the body, though it is interrupted, as shown in Fig. 2-4 and 5-3 by openings of glands and by various openings of the respiratory, digestive, and reproductive systems. The cuticle extends far into the gut and the tracheae of the respiratory system. Here the interruptions are internal.

***The Molting of Arthropods***

The arthropod exoskeleton suffers from one serious limitation. Once laid down, the chitinous plates do not grow, and the insect (or crustacean, such as a crab or lobster) can grow only by a process of *molting*, in which part of the old exoskeleton is shed. In the course of molting, the exoskeleton first becomes somewhat softened by a breakdown of the chitinous plates, presumably caused by some chemical activity of the layer of living cells just beneath. These cells then grow in size and sometimes undergo cell division as well, and then secrete the flexible parts of a new cuticle with folds. The remains of the old exoskeleton then split away from the living cells beneath and are shed, either in one piece or by splitting into several fragments. Finally, the animal expands to stretch the new cuticle before it hardens. The arthropod now has a new, enlarged, rigid skeleton (often somewhat compressed, telescoped, or pleated), which it gradually fills up as it grows. When the exoskeleton is completely filled so that growth is limited, the arthropod is ready for another molt. During the interval between the shedding of the old and the hardening of the new plates, the animal continues to be encased in a soft, flexible skin. Being soft bodied, it may be vulnerable both to predators and to injury from environmental influences such as drying. The "soft-shelled" crab is a normally well-armored arthropod that is temporarily in this embarrassingly defenseless stage after molting. Moderately large terrestrial arthropods, such as certain land crabs, are especially helpless while molting; if such an animal were as large as a dog, its soft body would probably collapse like a stranded jellyfish. There are behavioral changes that help protect these vulnerable animals. They often become seclusive, retreat into dark, moist, protected holes and forgo feeding until hardened. Not only may their mouth parts, pincers, and so forth, be too soft to capture or chew their food, but the animals cannot move efficiently or defend themselves against defensive action by their prey.

No one has yet discovered why arthropods must go through this dangerous phase of complete replacement of their all-important exoskeletons in several successive molts as they grow. Other animals have arrangements for efficient piecemeal growth, replacement, and even reorganization, involving changes in shape of their skeletons. Many insects achieve a considerable degree of breakdown, during molting, of the chitin-protein complex in their exoskeletons. The limitations imposed by the way in which an arthropod's exoskeleton is organized for growth seem to be major factors preventing the attainment of large body size, at least in terrestrial forms. In view of the frightful efficiency of insects and the serious nature of their competition with ourselves for food, we may be thankful for this handicap under which they operate.

### *Vertebrate Endoskeletons*

Vertebrate animals are the only ones that possess highly developed and efficient *internal* skeletons that are jointed and provide both stiffening and support for the body and at the same time permit movement of the parts thus strengthened. The aquatic vertebrates, of which the most abundant and the first to appear in geological history were the fishes, benefit principally by being able to attach rather sturdy muscles to a firm, yet jointed framework against which these muscles can pull. The length and shape of the body remains much the same during locomotion. Only the posture (relative positions of the different parts of the body) changes. Interestingly, the vertebrates have an elaborate system of *proprioceptive* sense organs in their muscles and joints, which inform their brain continually of the relative positions and tensions of the various parts so that they may move in a coordinated fashion, neither setting muscle against muscle nor overextending their joints. Proprioception also occurs in arthropods but has not yet been as well studied among other invertebrates. Presumably in an unjointed invertebrate such as an octopus (with many more "degrees of freedom" than a mammal or an insect) the proprioceptive system would be quite differently designed.

The swimming of a small fish is far more rapid and efficient than that of an aquatic worm of the same size; although other organ systems of the fish may also be superior to those of the worm, the internal skeleton and the neuromuscular exploitation of the skeleton are the major factors in the efficiency of such aquatic locomotion. Terrestrial animals of any size have an especially pressing need for skeletal support because their bodies can no longer depend on the buoyancy of water; hence it is on land that the internal skeleton of the vertebrates pays the largest dividends.

The earliest fishlike vertebrates known from fossils had bony, dermal armor as well as an internal skeleton. Nothing is known about their immediate ancestors, which probably had less easily fossilized skeletons. Their remote chordate ancestors exhibit only notochords. The notochord in primitive living vertebrates, such as the lamprey, is a rod of rather soft, special connective tissue, enclosed in a heavy, fibrous sheath. Such a fully developed notochord runs the full length of the animal in the midline, dorsal to the gut, and ventral to the central nervous system. In such an animal, it apparently prevents accordionlike contraction during movement. When the longitudinal muscles in the body wall contract the body will bend sideways.

In addition to bone, vertebrate skeletons often employ, in varying degrees, a material called *cartilage*, which is tough but slightly flexible and often quite elastic. The mechanical properties of cartilage can be judged by the simple experiment of bending one's own ear (it is stiffened by elastic cartilage). These three materials—bone, cartilage, and notochord—in varying proportions and with different arrangements of living cells and extracellular proteins and other ground materials make up the skeletons of all vertebrate animals. Bone may be totally absent; cartilage has a limited distribution in many adult vertebrates; the notochord, perhaps functionally significant only in lampreys and hagfish, is markedly reduced in reptiles, birds, and mammals.

If a bone is analysed chemically, one finds that after as much as possible of the soft tissue in and around it has been removed, its chief components are a complex of the salts $Ca_3(PO_4)_2$ and $CaCO_3$, laid down in a matrix of protein fibers. The proportions of the minerals themselves are close to those of the common mineral apatite, but the actual properties of bone are quite different from those of any homogeneous salt or mineral. For one thing, even the densest bone is not one solid piece. The inorganic material is continually being laid down and reabsorbed, modeled and remodeled, by living cells called *osteoblasts* and *osteoclasts.* Nourishing these cells are small blood vessels, which run through the bone much as conduits run through the ground. (See Fig. 3-3.) The center or core of bones, especially long bones such as those of the mammalian leg, is often filled with fatty tissue and/or *bone marrow*, consisting of cells that produce many of the red and white blood cells and the platelets. This marrow is not to be thought of as bone but as occupying space within the bone that is apparently not needed for structural strength. Because of the way bone is laid down and constantly remodeled, it is riddled by a continuous, interconnected series of small canals filled in life with living cells. At the boundaries of such canals, the cells living therein exchange materials with the blood and with the solid matrix in which they are embedded. The fibrous protein matrix

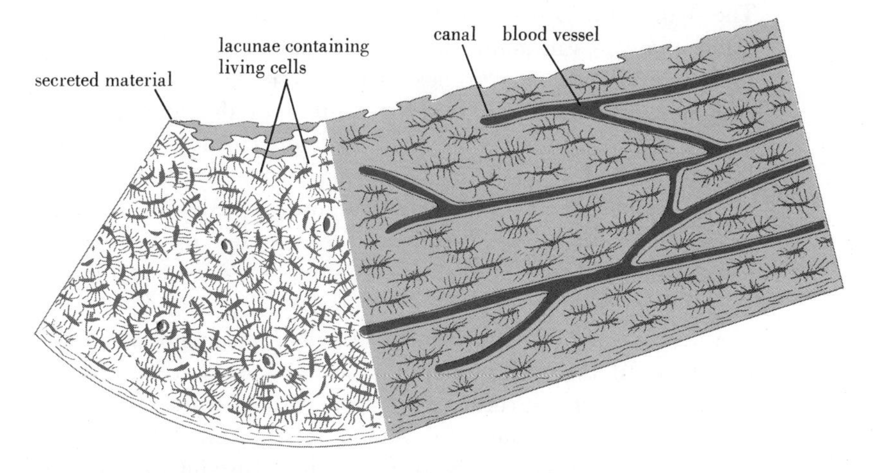


***Fig. 3-3*** *The microscopic structure of bone. Living cells along the lacunae, their metabolic needs supplied by blood vessels, are continually breaking down and secreting the hard material of the bone. The architecture is perpetually changing.*

in which the mineral salts are deposited is similar to the matrices secreted by other connective tissue cells, as described in the next section. Bone is a living, dynamic tissue. Vital changes are occurring continually. Bone also changes noticeably with age. A bone that was quite soft and flexible in youth may become hard and nearly solid in maturity. In old age, the minerals are often reabsorbed to a considerable degree leaving the bone soft and brittle and very susceptible to fractures or compression.

Cartilage is also the product of secretory activity by living cells that have surrounded themselves with a secreted matrix. It varies greatly in properties, ranging from very stiff to elastic and fibrous types. In the first, the cells are widely scattered singly or in small groups, and the secreted matrix is largely composed of a complex of protein and a modified carbohydrate called *chondroitin*. This substance, like chitin, contains units resembling glucose except for the replacement of an OH group by an $NHCOCH_3$ group or by having sulfate groups attached. At the other extreme of a series of different kinds of cartilage, the elastic type has a greatly increased proportion of elastic protein fibers. In two major groups of living vertebrates—the sharks and rays and their relatives and the cyclostomes (lampreys and hagfish)—the whole skeleton (excepting only the notochord) is composed of cartilage throughout life. In all other vertebrate classes, much of the skeleton is composed of cartilage during development and, indeed, in some forms ossification (conversion to bone) never occurs or does so only locally.

Among the vertebrates, some bones called *dermal* are laid down directly as bone and grow by remodeling and by addition to the edges. The scales of bony fish are dermal bone. These often fuse to make large armor plates, as in many catfish. But such bone also has been modified in time by evolutionary processes so as to make up most of the cranium, even in mammals, and part of the pectoral or shoulder girdle, especially among fish, amphibians, reptiles, and birds. All of the other bones of the body are of a type called *endochondral* or *replacement* bone, which is initially laid down as cartilage. At some genetically determined stage in development, such cartilages are invaded by blood vessels and osteoblasts and are replaced systematically by bone in whole or in part. In mammals, in the shaft or *diaphysis* of long bones such as the femur, a center of *ossification* spreads from an initial point near the middle until the entire diaphysis is converted. (See Fig. 3-4.) This process may take several years. In fact, it is not completed in man in some of the small bones of the hand and foot until the age of about 21. Meanwhile there is usually also a center of ossification in each of the two ends or *epiphyses* of the bone. These two centers of ossification also spread, but more slowly, all but meeting the diaphyseal bone. Between the epiphyseal bone and the diaphyseal bone, a thin plate of cartilage

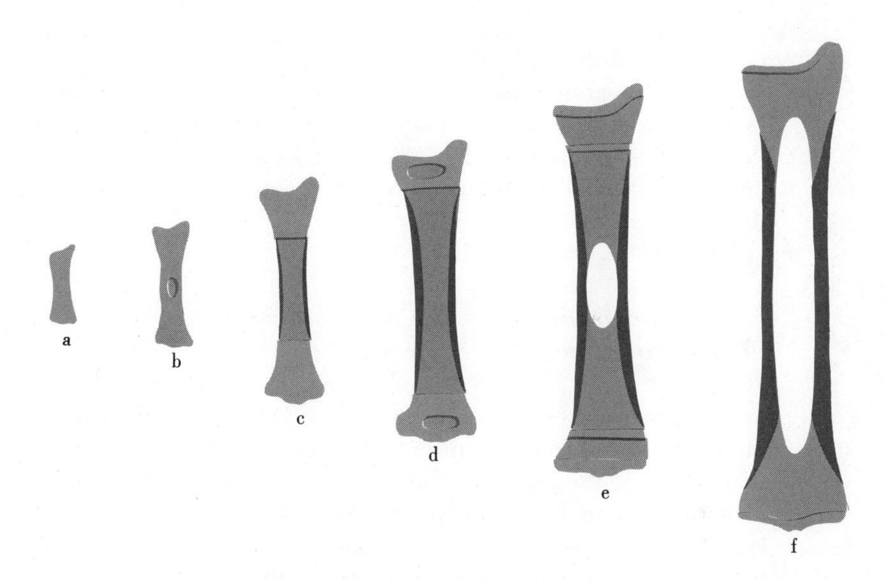


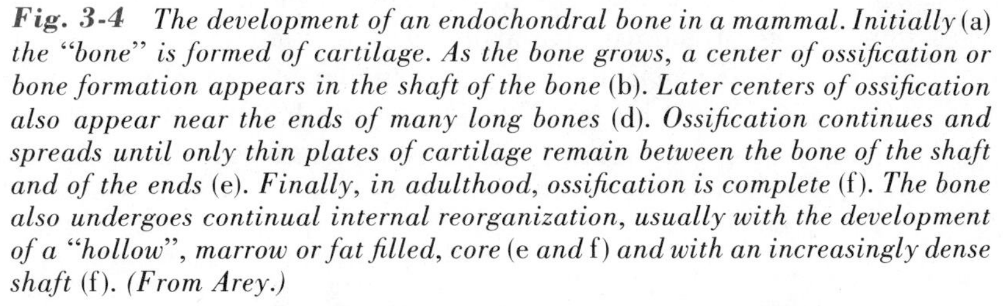
***Fig. 3-4*** *The development of an endochondral bone in a mammal. Initially* (a) *the "bone" is formed of cartilage. As the bone grows, a center of ossification or bone formation appears in the shaft of the bone* (b). *Later centers of ossification also appear near the ends of many long bones* (d). *Ossification continues and spreads until only thin plates of cartilage remain between the bone of the shaft and of the ends* (e). *Finally, in adulthood, ossification is complete* (f). *The bone also undergoes continual internal reorganization, usually with the development of a "hollow", marrow or fat filled, core* (e *and* f) *and with an increasingly dense shaft* (f). *(From Arey.)*

remains throughout the growth period of the animal, and it is here that elongation of the bone occurs; that is, growth, in the sense of elongation, occurs as cartilage deposition in the plate of cartilage between the epiphysis and the diaphysis. Growth in thickness occurs by constant remodeling of the surface and interior of the bone by the osteoblasts and osteoclasts. At the joint end of the epiphysis, a thin layer of specialized cartilage also remains and functions for life to line the joint space, presumably being designed for resilience and resistance to wear. The age of the animal can be calculated, up to adulthood, by a careful inspection of the degree of ossification of the endochondral bones and comparison with a series of known ages. This method is often used in medicolegal cases to estimate the age of an unidentified body. At full maturity, the diaphysis fuses with the epiphysis, eliminating the cartilaginous plate in between.

In reptiles, amphibains, and fish, ossification normally occurs at only a single center per bone, starting centrally and spreading throughout, so there is no distinction between diaphysis and epiphysis. Elongation occurs at the ends of such bones by remodeling and extension. In such animals growth is often indeterminate, occurring throughout life, though at a very slow rate in later years. Mammals all cease growing at the time the epiphyseal plates ossify. This mammalian development was probably associated evolutionarily with greater locomotor efficiency and the need for better joint design.

## The Organized Vertebrate Skeleton

The materials comprising vertebrate skeletons are formed into characteristic units, called "bones" for convenience—although, as noted previously, the skeletons may be composed of cartilage. Most prominent of the bones is the *skull*, made up of many individual bones that have fused together in adults, but most of which can be individually identified during development. (See Fig. 3-5.) The *brain case* and *palate* (subdivisions of the skull) contain both dermal and endochondral components whose homologies make up one of the most complex and beautiful stories of comparative anatomy. (See Fig. 3-6.) (Two structures are said to be homologous if they are similar in design, either in an adult or during some stage of development, that it is reasonable to conclude that one form has evolved from the other or that both forms have

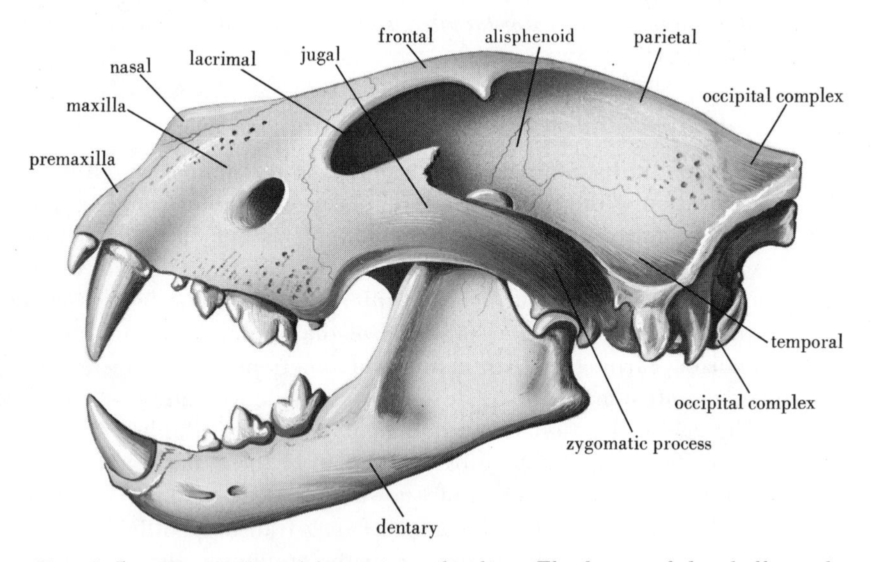


*Fig. 3-5 The skull and lower jaw of a lion. The bones of the skull can be traced to those of certain crossopterygian fishes and early amphibians such as are shown in Fig. 3-6. In the lion, the canine teeth are well developed for inflicting deep penetrating wounds while the carnassial teeth—the largest of the upper and lower jaw, respectively—are so arranged as to serve as shears for cutting through tough tissues such as muscle, tendon, or bone.*

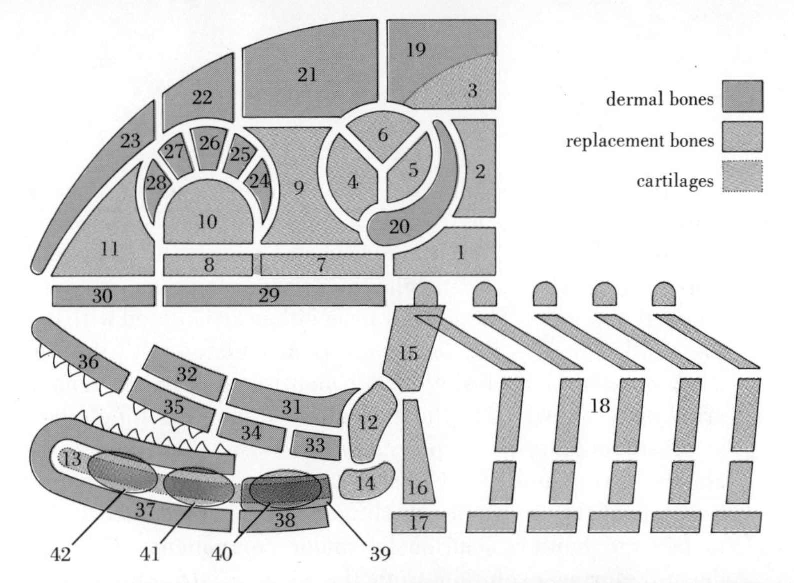


| | replacement bones and/or cartilages | | dermal bones | | |
|---|---|---|---|---|---|
| | **region** | **components** | **region** | **components** | |
| neurocranium | occipital region | 1. basioccipital<br>2. exoccipital<br>3. supraoccipital | rear-roof | 19. postparietal (interparietal) (supraoccipital) | |
| neurocranium | auditory region | 4. proötic<br>5. opisthotic<br>6. epiotic | cheek area | 20. squamosal | palatal series |
| neurocranium | basal plate | 7. basisphenoid<br>8. presphenoid | midroof | 21. parietal<br>22. frontal | 29. parasphenoid<br>30. vomer<br>31. pterygoid<br>32. palatine |
| neurocranium | orbitosphenoid | 9. alisphenoid<br>10. orbitosphenoid | circum-orbital series | 24. postorbital<br>25. postfrontal<br>26. supraorbital<br>27. prefrontal<br>28. lacrimal | |
| neurocranium | ethmoid region | 11. ethmoid complex | foreroof | 23. nasal | |
| splanchnocranium | visceral arch I (mandibular arch) | upper jaw<br>12. quadrate | maxillary series | 33. quadratojugal<br>34. jugal<br>35. maxilla<br>36. premaxilla | |
| splanchnocranium | | lower jaw<br>13. meckel's cartilage<br>14. articular | mandibular series | 37. dentary<br>38. angular<br>39. surangular<br>40. splenial<br>41. prearticular<br>42. coronoid | |
| splanchnocranium | visceral arch II (hyoid arch) | 15. hyomandibula<br>16. ceratohyal<br>17. bashihyal | | | |
| splanchnocranium | visceral arches III-VII (branchial arches) | 18. branchials | | | |

***Fig. 3-6*** *A diagrammatic representation of the bones of the primitive tetrapod skull, lower jaw, and gill arches. Many of those which persist in mammals can be identified in Fig. 3-5. (After Rand and Torrey,* Morphogenesis of the Vertebrates, *John Wiley & Sons, 1967.)*

evolved from a common ancestor. The evidence is usually not only from the anatomy of living species but from fossils as well.) The major functions of the skull are to enclose and protect the brain and the major sense organs, the eyes, the olfactory tissue, and the inner ears, and their respective nerves. The *upper jaw* is either articulated with or fused with the skull, whereas the *lower jaw* is articulated with the upper jaw to allow opening and closing of the mouth and other movements such as grinding or chewing. In the cyclostomes, jaws are totally lacking. Jaws serve, of course, as emplacements for the teeth and for controlling access to the mouth and pharynx. Jaws also permit manipulation and handling of the environment. Hinged jaws enabled ancestral vertebrates to become hunters. Curiously, major components of both jaws were derived during evolution from the skeletal structures supporting the gill arches of early fishes. Ultimately some of these gill-arch elements, converted for a while to jaw function, were destined to become tiny bones in the middle ear serving to conduct sound from the eardrum to the inner ear.

The next most important component of the skeleton is the *vertebral column*, or backbone, formed by a number of separate *vertebrae*, arranged in a series articulating with the skull anteriorly, and then with each other and with the ribs. (See Fig. 3-7.) A spool-shaped body or *centrum* of each vertebra usually surrounds or replaces the notochord locally. Dorsal to the centrum, and resting on it, is an arch of bone, the *neural arch*. The series of neural arches enclose and protect the spinal cord. Ventrally, the vertebrae may have *haemal arches* protecting important blood vessels. The vertebral column supplies substantial support for the body while yet allowing great flexibility. Flexibility can be limited in some directions or in some regions of the body by extra vertebral processes or by bony fusion. The skull and vertebrae together,

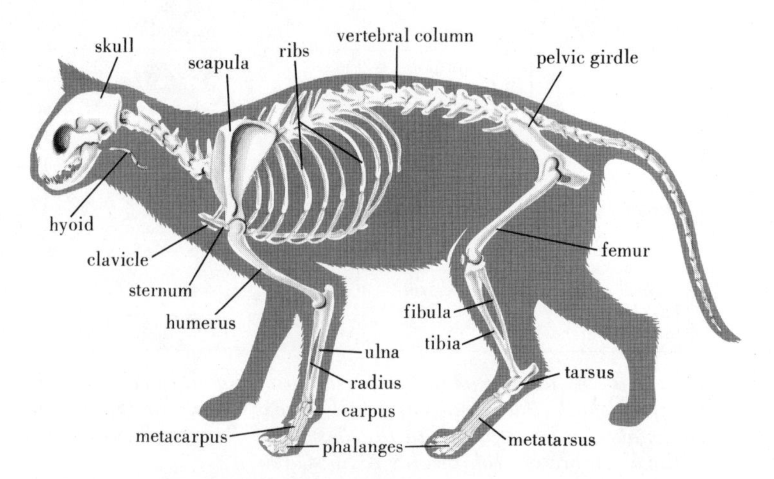


***Fig. 3-7*** *The skeleton of a domestic cat.*

then, enclose and protect the entire central nervous system in vertebrates above the level of the cyclostomes. Primitively, each vertebra is associated with at least one and often two *ribs* on each side and such a multiplicity of ribs still characterises the bony fish, the ribs making up the sharp bones that plague the gourmet and other predators. In mammals, the ribs are normally retained only in the *thoracic* region where they give shape and rigidity to the chest and serve as points of origin and insertion for the muscles that expand the chest in respiratory movements. In the *sacral* region in mammals, modified ribs serve to fuse the pelvic girdle to the sacral vertebrae. The last part of the vertebral column makes up the skeleton of the tail.

In the *pharyngeal* region of fish, a series of bones support the gills. These bones, often called the *visceral* skeleton, have had a varied fate in vertebrate evolution. As previously noted, they gave rise to portions of both jaws and to the *auditory ossicles* of tetrapods, especially mammals. They have also come to provide the *hyoid apparatus*, which gives skeletal support to the tongue, and the skeletal structure of the larynx and trachea. The skull, vertebral column, ribs, and visceral skeleton—along with such specialized bones as those in the *sternum*—are often grouped together as the *axial* skeleton.

In most vertebrates except the cyclostomes, their extinct relatives, and another class of extinct fishes called placoderms, there are paired *limbs* or *fins* associated with *pectoral* and *pelvic girdles*. Fig. 3-8 shows an example of the skeleton of the pectoral girdle and fin of an extinct bony fish ancestral to the amphibians and higher vertebrates. The simi-

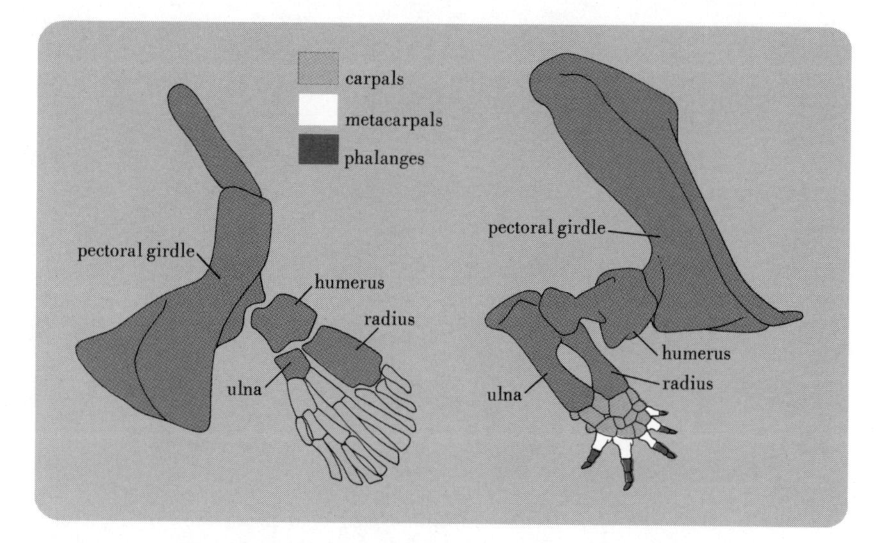


**Fig. 3-8** *The pectoral girdle and limb (fin) of* (left) *a crossopterygian fish and* (right) *a primitive amphibian. Note the basic similarity in limb pattern. (After Gregory and Romer,* The Vertebrate Body, *W. B. Saunders Co., 1962.)*

larity between these fins and typical tetrapod limbs is evident.

Basically, speaking now of mammals, the pectoral girdle serves as an anchor for the *arms* or *forelegs*. The pectoral girdle of amphibians, reptiles, and birds is a rather complex and frequently rearranged array of both dermal and endochondral bones. Among mammals the *scapula*, very much enlarged, and the *clavicle* persist. Muscles connect the pectoral girdle to the vertebrae, the ribs, the skull, and the sternum as well as to the bone of the upper arm, the *humerus*. These muscles move and/or stabilize the arms and shoulders. The arm itself consists of a single, large proximal bone, the humerus, usually with a rather mobile, ball-and-socket joint with the scapula of the pectoral girdle. (See Figs. 3-7, 3-9, 3-10.) The humerus articulates distally, at the *elbow*, with two bones lying side by side, the *radius* and the *ulna*. These in turn articulate at the *wrist* with the small bones of the *hand*, the *carpals*. Beyond these are the digits, consisting each of a proximal *metacarpal* and distal *phalanges*. In mammals, there are often five digits and, frequently, as in man, the first of these, the thumb, has two phalanges and the others have three phalanges each. There is no direct bony connection between the pectoral girdle and the vertebral column in most mammals. The clavicle (a portion of the pectoral girdle) in many mammals does articulate with the sternum, which in turn articulates with the ribs.

The pelvic girdle in mammals consists of three separate elements on each side, the *ilium*, *ischium*, and *pubis*, which fuse and make a

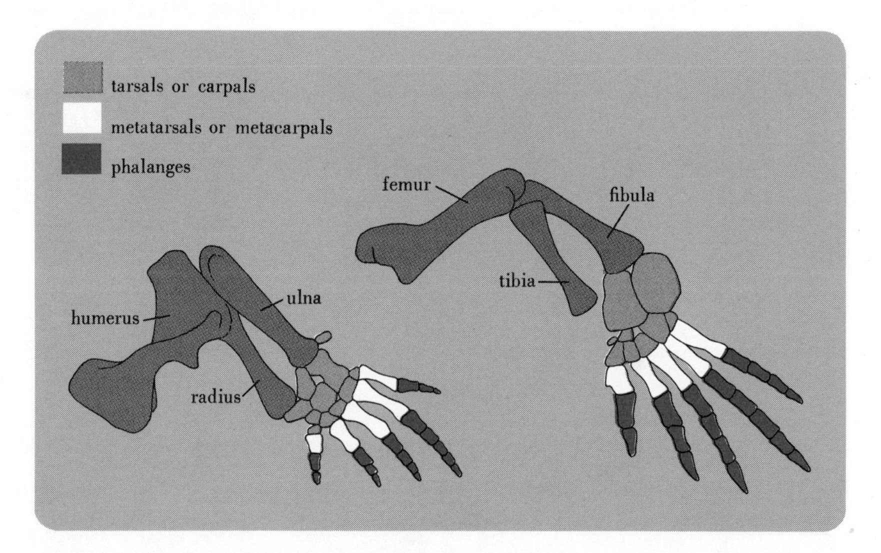


***Fig. 3-9*** *The left front and hind limbs of a primitive reptile, showing how limbs were patterned in early tetapods. Note the basic similarity of design of the two limbs. (After Romer,* The Vertebrate Body, *W. B. Saunders Co., 1962.)*

complete bony ring in adults. All six of the bony components are solidly fused with each other and also, dorsally, with the sacral vertebrae and their highly modified ribs. Ventrally the bones of the two sides are fused together in a generally immovable joint called the pubic and ischial *symphysis*. It is through this bony ring that mammalian young must pass at birth. Compromise in design is necessary to meet the needs to provide for sturdiness of the female pelvic girdle for weight-bearing and locomotion, and still provide sufficient diameter for passing the young. The size limitations of the bony pelvic girdle also require that there be a compromise between maximal maturity and size of the young at birth (for greater self-sufficiency) and smaller size or greater flexibility of the fetal parts to allow easy delivery. In some mammals, the joints of the pelvic girdle soften, under endocrine control, and become elastic as *parturition* (delivery of the young) nears.

The large, proximal bone of the leg, the *femur*, articulates with the pelvic girdle and in turn articulates at the *knee* with two bones lying side by side, the *tibia* and the *fibula*. (See Figs. 3-7 and 3-9.) These articulate at the *ankle* with the small bones of the *foot*, the *tarsals*. There are often five digits, each with a proximal *metatarsal* and several phalanges. The number of *toes* and phalanges is often the same as for the hand. There are also many remarkable variations and specializations of the limbs and girdles, some of which are shown in Fig. 3-10.

Conservation of the basic limb structure in vertebrates implies

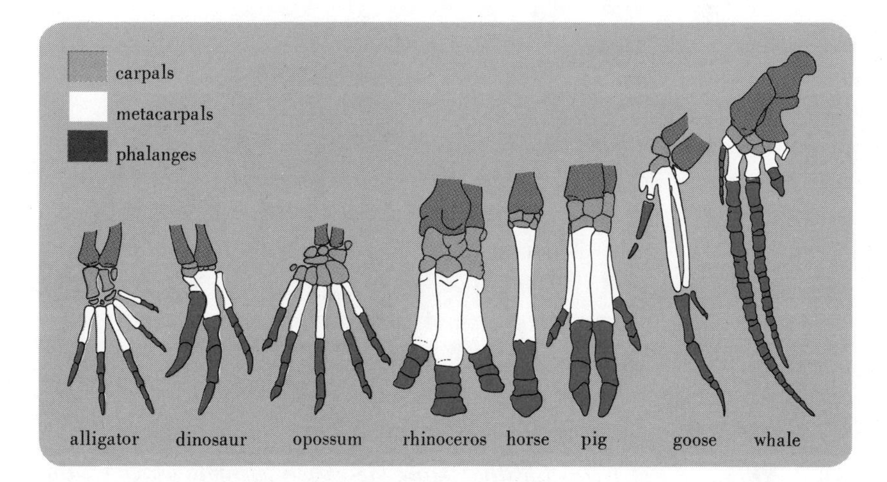


***Fig. 3-10*** *A selection of specialized tetrapod hands. The first finger (when present) appears to the left. The carpal elements are often reduced in number or fused. The number of digits, their length and proportions, varies widely. (After Romer.)*

strong functional advantages, but limb structure is remarkably varied as well and, of course, function is also strikingly varied—often specialized, often versatile. Even more impressive evolutionary versatility is exemplified, however, by arthropod limbs. (See Fig. 3-11.) Among vertebrates, limbs have been designed or redesigned for swimming, walking, burrowing and digging, climbing, brachiating, flying and gliding, jumping, and an almost endless list of variations of these functions. Limbs have also been exploited as weapons using claws or hooves, as extension poles for sensory probing of distant or partly hidden objects (such as insects in holes), as organs for examining objects and for bringing them to the mouth, or for carrying, for grooming, or even for gaffing fish. They may even be used, in effect, as stilts. These specializations are associated with structural rearrangements. Not surprisingly, the length and bulk of the longer bones is often modified and the joints are often redesigned to facilitate particularly necessary or desirable movements or to reduce the likelihood of injury. Bones have often been eliminated in whole or in part or fused with adjacent bones. The carpals, tarsals, and digits are often reduced in number or modified. In a few cases such as in whale forelimbs, the number of carpal or digital bones may be increased. (See Fig. 3-10.) In fact, limbs have been completely lost in several groups, independently, such as the apodan amphibians, many snakes, and some lizards. In whales, the hind limbs are lost. Many functions must be either lost or replaced by adaptations of other parts of the body in such limbless organisms: locomotion, the capture and handling of

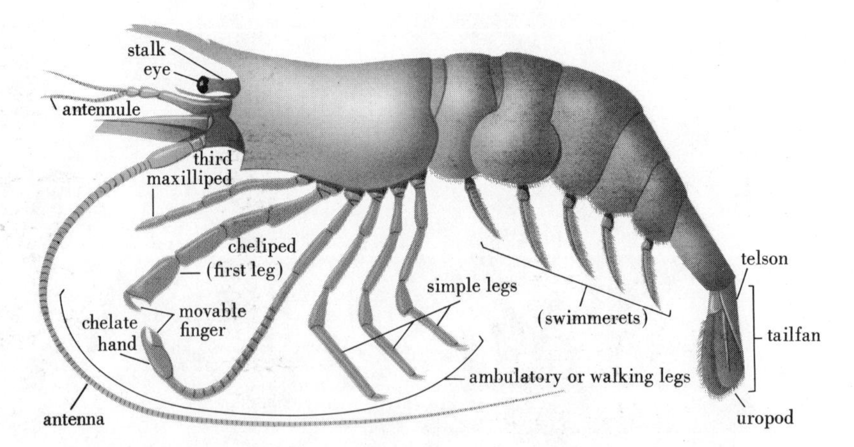


*Fig. 3-11 A crayfish illustrating the variety of limbs which characterize an individual arthropod. A startling variety of limb structures and adaptations occur in this phylum.*

food, grooming, behavioral posturing as in courtship and mating, and so forth. Viewed in this light, the common snake diet of bird's eggs may not be surprising. Eggs are not only unable to escape or to counter-attack but are a relatively easy package to swallow without much handling.

An interesting case of limb modification in mammals that has been well studied has been the horse, where the wrist and ankle joints are substantially redesigned, the carpals and tarsals reduced, and the digits reduced to a single one for each foot. The horse's hoof itself is a much modified nail on the terminal phalanx of the third digit. Horses are apparently highly specialized for swift, long-distance movement across such terrain as prairies. Their ankles are particularly highly modified for stability. Their limbs, however, have been restricted in their freedom of movement and are useful, principally, only in walking or running; grooming, handling capacity, and other functions have been sacrificed.

## CONNECTIVE TISSUE

A skeleton is more than a set of individual bones and cartilages; indeed, all of the individual parts and organs of any animal would fall apart were it not for the connective tissue consisting of cells and the inert secreted materials in which they are embedded. The term *tissue* is used here and elsewhere to denote a more or less continuous set of cells and their adjacent secretion products, which act together for some definite purpose. In almost all multicellular animals, a variety of cells take part in this important function of connecting one part of the body with another. Usually these connective tissue cells are rather elongated and their cytoplasm may form radiating branches that can change in shape slowly over hours or days. Close around them, these cells secrete fibrous, gelatinous, and sticky, mucoid materials to form sheets, strings, or masses of living tissue that serve in joining and linking parts of the body. The class of proteins known as *collagen* are among the most important of these secreted materials. Collagen exists in the form of long fibers and has a well-defined ultrastructure enabling it to be recognized in electron microscope pictures of many types of living material.

The thicker and tougher deep layers of skin, in some vertebrates, are a type of connective tissue in which insoluble fibrous proteins predominate. Equally important are the many membranes that line various cavities in the bodies of animals, such as the *peritoneal* lining of the abdominal cavity or the *pleural* lining of the thorax in mammals.

Such membranes, among the vertebrates, may also serve to suspend organs within the body cavities. Thus, a dorsal and a ventral *mesentery* suspend the vertebrate gut from the mid-dorsal line and the midventral line longitudinally. There are many cordlike or sheetlike *ligaments* connecting bones to one another and *tendons* connecting muscles to bones. Much thinner sheets of loose and slippery connective tissue enclose and subdivide each muscle, allowing either movement of the muscle as a whole or independent movement of its parts. Nerves are similarly enclosed in connective tissue sheaths, which allow other tissues, such as nearby muscles, to move without tugging at the nervous tissue. In fact, wherever organs come together, be they bones or muscles, nerves or blood vessels, it can be assumed that a specialized kind of connective tissue is making the junction and is providing the needed degree of solidity or flexibility.

Connective tissue also fills in chinks and spaces in various parts of animal bodies. One special type consists chiefly of cells containing large numbers of fat globules in the cytoplasm. Such stored fat is drawn upon in times of need, especially when metabolic demands are very high as during a reproductive period or when food is scarce or when weather conditions preclude the gathering of food. Small mammals that hibernate, such as the big brown bat or the meadow jumping mouse, store as fat as much as 100 percent of their summer body weight in the fall of the year to provide fuel sufficient for their metabolic needs during the winter fast. Curiously, the chemical makeup of the fat that is stored varies from species to species. In many mammals during hibernation, the fat is more unsaturated than fat stored during the summer. A parallel situation may be seen in nonhibernating, arctic mammals whose legs are subjected to severe winter temperatures. In the moose, the fats in the distal parts of the limbs are more highly unsaturated (and with a lower melting point) than the fats in proximal depots. The apparent significance of such unsaturated fat storage is to preserve liquidity under cold conditions. Fatty connective tissue is also important as a soft padding and filler between many organs. Such a layer surrounds the kidneys in cattle and can be viewed easily in the butcher shop. Fat also gives characteristic contours to the body, often with sexual differences, as in our own species. Finally, a special sort of fatty tissue known as *brown fat* is widely known among mammals that hibernate. Such tissue has recently been shown to generate heat quickly and in large quantity when the animal is awakening. The mechanism of such heat generation and how it is triggered remains obscure. Previously muscle action, especially shivering, was thought to be chiefly responsible for such warming up.

Thus, although connective tissue may be stuffing and packaging for the "real" organs, it has many additional refined functions. Millions of relatively simple cells scattered throughout the body are capable of secreting a wide variety of materials differing in mechanical properties; yet each group of connective tissue cells does in fact secrete only the correct materials and in the appropriate amounts. Hundreds of different mechanical needs are filled by these versatile cells and the substances which they secrete. Hence we must conclude that despite its lack of such visibly complex structures as are found in bone, connective tissue is far from being random stuffing.

## FURTHER READING

Barrington, E. J. W., *Invertebrate Structure and Function.* London: Nelson, 1967.

Bloom, W., and D. W. Fawcett, *A Textbook of Histology*, 9th ed. Philadelphia: Saunders, 1968.

Cott, H. B., *Adaptive Coloration in Animals.* London: Methuen, 1957.

Fawcett, D. W., *An Atlas of Fine Structure: The Cell, Its Organelles, and Inclusions.* Philadelphia: Saunders, 1966.

Marples, M. J., "Life on the Human Skin," *Scientific American*, 220 (1): 108–115, 1969.

Montagna, W., *The Structure and Function of Skin*, 2d ed. New York: Academic Press. 1962.

———, "The Skin," *Scientific American*, 212 (2): 56–66, 1965.

Portmann, A., *Animal Camouflage.* Ann Arbor: University of Michigan Press, 1959.

Romer, A. S., *The Vertebrate Body*, 3d ed. Philadelphia: Saunders, 1962.

———, *Vertebrate Paleontology.* Chicago: University of Chicago Press, 1966.

Schubert, M. and D. Hammerman, *A Primer on Connective Tissue.* Philadelphia: Lea and Febiger, 1968.

*chapter* **4**

# *Digestive Systems*

Securing and processing food is obviously important to animals. The ease or difficulty of the task, of course, varies enormously according to the animal's way of life. Near one extreme of nutritional security are intestinal parasites, for which the host animal does all the work of catching the food and most or all of the digesting as well. In consequence, the most highly organized of these intestinal parasites, the tapeworms, have no digestive system: the digestive products of food molecules such as sugars simply diffuse from the intestinal contents of the host into the body of the tapeworm. Blood parasites, such as the organisms causing malaria, are even more extreme examples of nutritional security. Needless to say, they may also be provided with a controlled environment in terms of such factors as temperature, pH, and oxygen tension. Their metabolic waste products are even cared for by the host. The parasitic way of life, however, is not entirely free of care. The needs of the parasite must often be subordinated to those of the host, lest the individual host be killed or the host species be made noncompetitive. Parasites must also continually evolve as their hosts develop new counterweapons to infestation. The problems of reproduction and the colonizing of new hosts are enormous. Some parasites surely set the record for astronomical numbers of offspring produced in order to insure continuation of the species. In the common large roundworm, *Ascaris*, each adult female produces about 200,000 eggs per day. Such reproduction is, of course, also at the expense of the host. Most animals, however, must work for their own living, and much of their immediate business is securing food and chemically reducing it to small molecules that can move into their bodies.

Many marine animals have recently been shown to be able to absorb actively various organic compounds, such as amino acids, which are present in sea water in very minute quantities. This ability may be very important in animals that live in mud and sediment, where much decay of organic material is taking place. Perhaps it also explains the ability of the curious group of marine worms, known as the Pogonophora, to survive even though they have no digestive systems.

The oceans, and many bodies of fresh water, also contain a wealth of small plants and animals, referred to as *plankton*. Many different kinds of animals—sponges, clams, small crustaceans, and numerous others—live by filtering plankton out of such waters. Such *filter feeders* are often sessile, remaining attached to a surface and using muscular or ciliary pumps to draw in water and trap the plankton. Many of them secrete mucus to assist in trapping the smallest plankton.

Protozoans engulf small food particles by surrounding them with a membrane similar to the cell surface membrane, thus forming *food vacuoles*, which carry the food particles within the cytoplasm. Similar engulfing of particles also occurs in many cells of multicellular animals. In coelenterates and many other invertebrate animals, cells emerge from the wall of the digestive cavity, move about in its contents, and engulf food particles. They then return to the animal's tissue, move about among the cells, and presumably give off some products of digestion to these other cells. The details of this process have not yet been determined. Such wandering cells are much like a common protozoan, the amoeba, which has a rather fluid protoplasm and changes its shape from moment to moment by extending temporary projections or by contracting into a nearly spherical form. Hence these cells are called *amoebocytes*. In our own bodies and those of all other animals, cells of this general type circulate in the blood and tissue fluids, moving about through the spaces among other cells and often engulfing such foreign particles as bacteria. They are called *phagocytes* (including most types of white blood cells) and form part of our defense against foreign bodies and infectious organisms.

When food is digested inside a single cell, the process occurs in a vacuole or fluid-filled cavity in the cytoplasm. Such compartmentalization helps prevent self-digestion. As far as we know, the essential steps in the process of digestion are the same, whether they occur in a food vacuole in an amoebocyte or in the large and visibly specialized digestive tract of a mammal. These steps have been thoroughly studied in many multicellular animals; we can best consider them in animals in which division of labor has segregated the major processes into separate locations.

All animals more specialized than the flatworms have a tubular digestive tract leading from an anterior *mouth* to a posterior opening, the *anus*, through which food residues or *feces* are eliminated. There is an obvious efficiency to a system for digestion that is subdivided into sequentially specialized stages, each carried out by an organ, such as the mouth, *stomach*, or *intestine*, that is suited for a particular process. (See Fig. 4-1.)

## DIVISION OF LABOR ALONG THE DIGESTIVE TRACT

Digestion often begins with mechanical fragmentation of the food, such as by chewing. (See Fig. 4-2.) A wide variety of toothlike structures are used by different kinds of animals both to seize food initially and to reduce it to small particles. In birds, however, most of the reduction is accomplished in a modified part of the stomach, the *gizzard;* birds have no bony teeth and the horny bill can cut food, at best, into only slightly smaller pieces than the original. The gizzard is a muscular chamber that squeezes and kneads the food. In most birds, small stones or grit are swallowed and held in this chamber to aid in the grinding process. Eventually the pebbles are worn smooth and reduced in size, until finally they pass along through the digestive tube and must be replaced; therefore, seed-eating birds such as chickens may starve to death even though well provided with food unless they have a supply of grit. Some herbivorous dinosaurs, closely related to bird ancestors, apparently used gizzard stones in a similar fashion, for large, smooth stones, often fist-size or larger, are found in compact clusters in the midst of fossilized dinosaur bones. Dinosaur gizzard stones are so common, indeed, that they can be purchased inexpensively from several biological supply houses. Gizzards are found in another surviving group of dinosaur

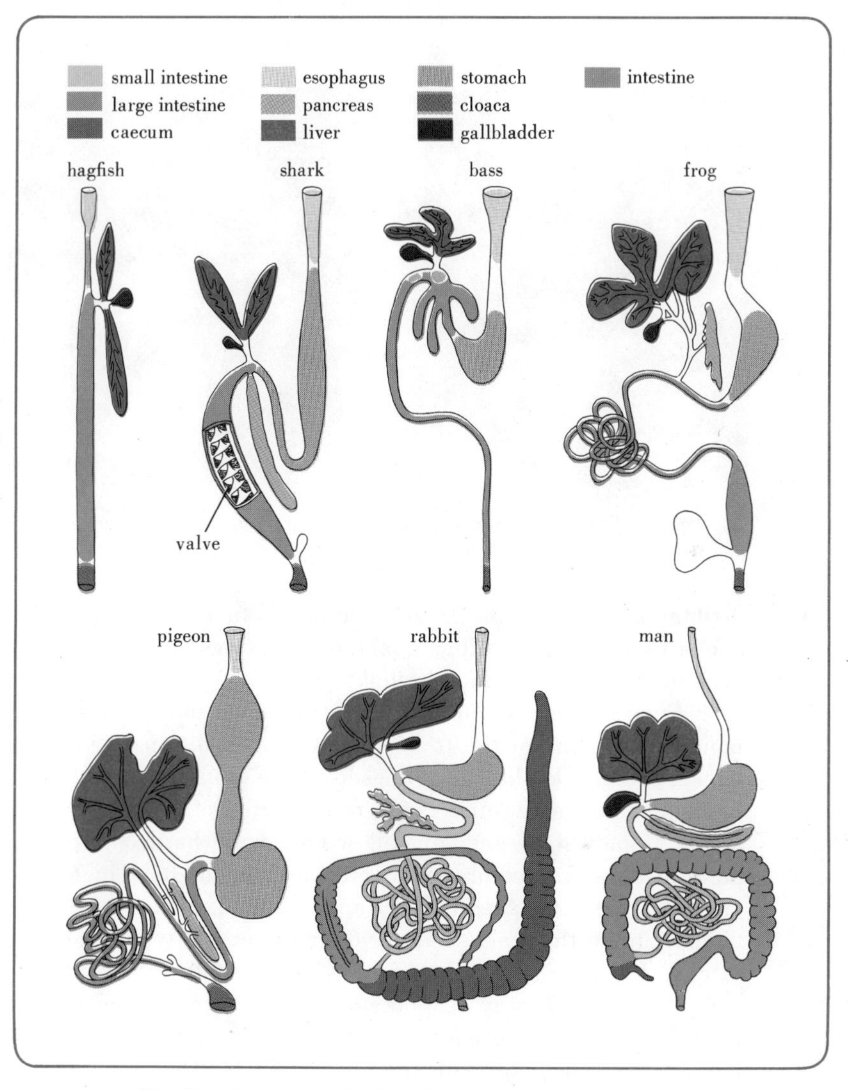


***Fig. 4-1*** *The digestive tract of selected vertebrates. (After E. Florey,* Introduction to General and Comparative Physiology, *W. B. Saunders Co., 1966, and after W. Stempell,* Zoologie in Grundriss, *G. Borntraeger, Berlin, 1926.)*

relatives, the crocodiles. Some other animals, including earthworms and some kinds of insects, employ internal mechanical fragmentation in their stomach as well. In almost all animals, some muscular churning supplements chemical digestion. Chemical attack occurs at the surface of food particles. The rate of chemical digestion depends in large part on the breaking up of large pieces into a number of smaller ones with a greater total surface area. Among bats, for example, insect prey are sometimes so fragmented by chewing that by the time they reach the stomach they cannot even be identified as to family. Digestion is so

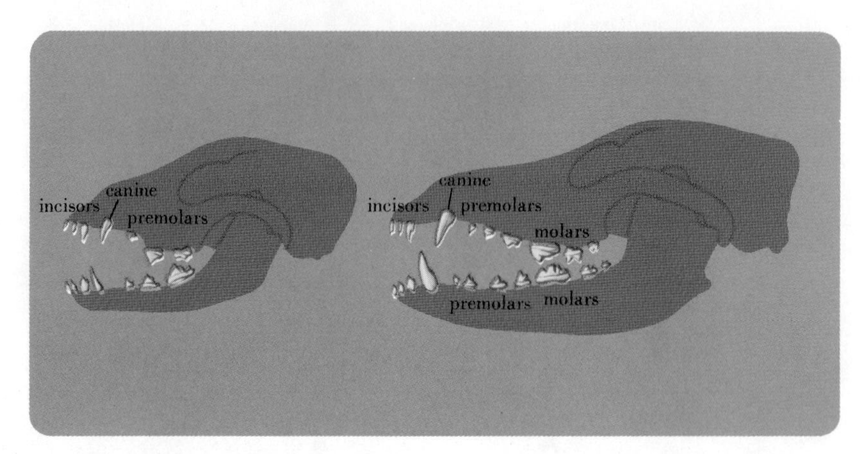


***Fig. 4-2*** *The deciduous or milk teeth of a puppy* (left) *and of an adult dog* (right).

facilitated by such fine fragmentation as to be largely completed in one or two hours, a significant metabolic savings for an animal that must bear the weight of its food in flight.

Even flatworms and coelenterates have a more or less specialized mouth, and many of the free-living, as opposed to parasitic, flatworms have a rather complicated proboscis at the mouth, which they push out to engulf food and at other times retract into the digestive cavity. Many of the roundworms and annelid worms have hard, sharp, or barbed teeth at the entrance of their mouth; arthropods have highly elaborate jaws and teeth for seizing prey and breaking up animal or yegetable food. Many of the mollusks have a characteristic type of rasping structure, called the *radula*, that is moved back and forth by muscles and scrapes and files away at food sources even though these may be extremely hard and resistant. Certain snails, for example, may drill (or scrape) a hole through the hard shell of other mollusks, such as oysters, in order to eat the soft parts of the prey. The mechanical drilling is facilitated by chemical digestion of the shell. The prey animal is often much larger than the attacking snail. A few mollusks carry this process so far that the radula is used to dig burrows into wood or even rock. These can rapidly destroy piers or wooden ships.

In mammals, the *saliva* contains a *digestive enzyme* that begins the chemical digestion of starch to maltose in the mouth. There are instances, among invertebrates, in which food is digested or partly digested before being taken into the mouth. Spiders, for example, may inject a liquefying and digesting complex into their prey and later ingest the fluid products. In all animals with a "through-tract" digestive tube, food is moved from the mouth cavity into a stomach that is usually enlarged to form a chamber where food can be stored temporarily and also undergo chemical digestion. At the posterior end of the stomach there is almost always a muscular valve called a *sphincter*, that is, a narrow part of the digestive tube that can be closed by the action of muscles arranged in a ring around the tube.

The next section of the digestive tract is the intestine. This is the portion, in all animals, in which the bulk of digestion occurs and from which almost all of the digested food molecules are absorbed into the body. The material that enters the intestine from the stomach is largely liquid; any remaining solid particles are quite small and may be indigestible residues that will be passed out in the feces. Digestive glands are very prominent in or near the walls of the intestine, and their products are of great importance in completing the processes of chemical digestion. In animals with well-developed circulatory systems, there is a rich blood supply to the walls of the intestine in order that the food molecules absorbed from the cavity or *lumen* of the intestine can be carried off.

In terrestrial animals, the more posterior parts of the digestive tract are often specialized for the absorption of water, so that the contents are made drier and harder. Such water as is reabsorbed has often been secreted into the lumen of the digestive tract higher up in order to lubricate the food, to facilitate its movement, to carry the small particles, and to dissolve the products of digestion that are absorbed only when in solution. Were the water not conserved by reabsorption, terrestrial animals, in particular, would suffer a severe loss. The ionic content of the digestive secretions is also normally reabsorbed; again this is a necessary conservation. Often another feature of this part of the digestive tract is the large population of bacteria and other microorganisms which thrive in such a stable and rich environment. Many animals have branches from the large intestine called *caeca* (singular caecum) and food residues may remain there for some time undergoing chemical changes caused by the microorganisms. (See Fig. 2-7.) Finally, there is often a *rectum*, or more muscular chamber, where the feces accumulate and are periodically extruded through the anus. Muscular sphincters regulate the anal opening.

Animals that feed largely on other animals, and hence obtain much of their fuel from molecules much like those in their own bodies, usually have relatively short and simple digestive tracts with only small caeca, if any. *Herbivorous* animals usually have a harder task of chemical digestion and must hold food under digestion for longer periods and in more elaborately subdivided stomachs or large caeca, where the food material may be acted upon by microbes. Such microbes are often able to digest cellulose to sugars which may then be utilized both by themselves and by the host animal. In ruminants—such as cattle, sheep, and antelopes, for example—grass or other herbaceous material is cropped and on being swallowed passes into a large subdivision of the stomach called the *rumen.* The rumen acts as a fermentation chamber. Billions of anaerobic bacteria and protozoans live and multiply there, many specifically adapted to this environment. The microbes are provided with an oxygen-free, warm, constant-temperature, ion-rich, nutrient broth with a regulated pH. The food is even regurgitated when the host is at rest and rechewed (chewing the cud) to increase fragmentation. The microbes in such a broth use, reduce, and convert not only the cell contents of the grass but also the cellulose of the cell walls to small, utilizable molecules, ultimately largely short-chain fatty acids and amino acids. The amino acids normally come from the digestion of cell proteins. They can, however, be synthesized by these microbes if a nitrogen source is provided. Recently urea has become a common component of cattle feed in order to provide such a nitrogen source in a very cheap form. Cattle can now be maintained on feed such as straw containing almost no protein if an appropriate daily ration of urea is included. The products of digestion not only nourish the microbes, themselves, but are passed on through the stomach and absorbed in the intestine of the host for the host's nutrition as well. In the absence of such a microbial reaction chamber, the energy contained in the cellulose would have to be foregone by mammals, none of which have enzymes to digest it. The host eventually recovers additional benefits from the food utilized by the bacteria and protozoans since the heat produced can be used to maintain the host's body temperature. The microbes eventually die and their bodies may be digested and absorbed.

In rabbits, such herbaceous food is partly processed, probably in a similar fashion, in the caecum. Since the products of digestion are beyond the absorptive part of the mammalian intestine, this nourishing, processed material is passed through the anus and re-eaten, a process called *refection.* Indeed, a rabbit's nutrition *depends* on refection. Genuine fecal material is passed separately, is readily distinguishable in texture and content, and is not refected.

Even in man, the bacterial flora of the large intestine are apparently necessary for the production of vitamin K. This vitamin must often be provided by dietary supplement when the intestinal flora are destroyed as they may be by massive doses of some antibiotics.

Digestive tracts in more highly specialized animals, such as mammals, often have folds and inward projections that serve to increase the surface area through which food may be transferred into the blood. The most important of these surface-increasing structures are the *villi,* which are found in enormous numbers on the inner surface of the mammalian small intestine. Each villus is a finger-shaped projection into the lumen from the internal surface of the intestinal tube; in the interior of the villus are blood and lymph vessels into which the products of digestion are transferred by a process discussed below, generally designated as *absorption*. (See Fig. 4-3.) Owing to the presence of villi, the total surface area for absorption is many times what it would otherwise be, and a given length of mammalian intestine can absorb much more food than it could were it a simple tube. In addition, the surfaces of the cells forming the innermost surface of the intestine are folded into *microvilli*, which are clearly visible only with the aid of the electron microscope. These minute projections also serve to increase the surface area available for the processes that enable these cells to move digested food products from the contents of the intestine to the blood and lymph vessels that will carry them to other parts of the body.

*Digestive glands* are essential in many parts of all digestive systems. They vary from single cells, which line the digestive tract and discharge *mucus* or digestive enzymes directly into the cavity, to cells formed into large separate organs. These organs discharge their secretions into ducts that may be, in turn, connected to other ducts to form tubules draining eventually into the cavity of the digestive tract. The large digestive glands of vertebrate animals are the *pancreas* and the

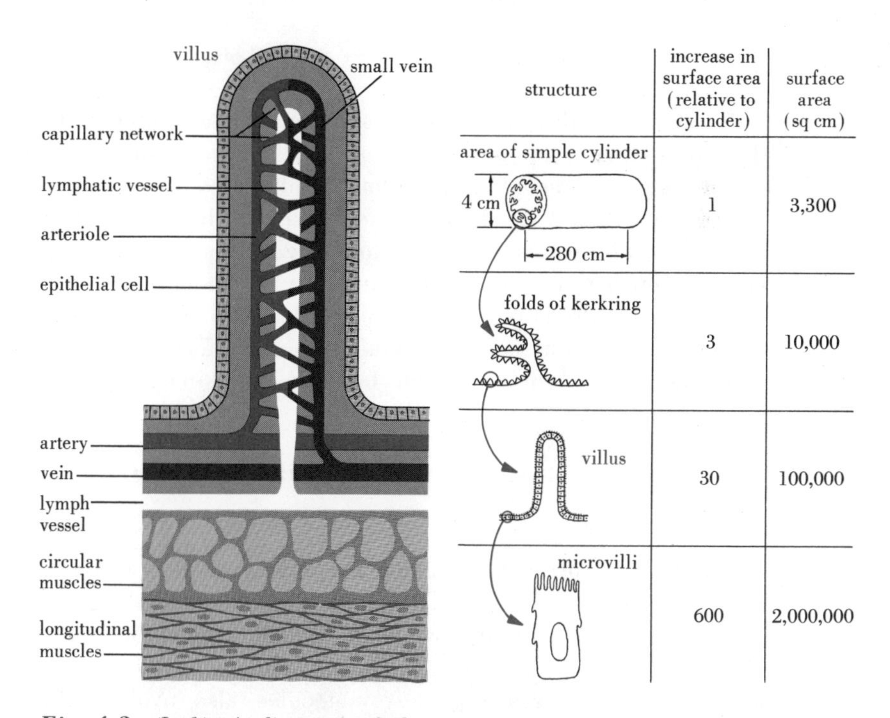


| structure | increase in surface area (relative to cylinder) | surface area (sq cm) |
|---|---|---|
| area of simple cylinder | 1 | 3,300 |
| folds of kerkring | 3 | 10,000 |
| villus | 30 | 100,000 |
| microvilli | 600 | 2,000,000 |

***Fig. 4-3*** (Left) *A diagram of the microscopic structure of an intestinal villus.* (Right) *A table of the structures by which the surface area of the mammalian intestine is increased. (After T. H. Wilson,* Intestinal Absorption, *W. B. Saunders Co., 1962.)*

*liver*, both of which drain into the small intestine. The liver is not known to secrete any digestive enzymes but does produce *bile*.

Food is needed by animals for three reasons: to provide energy; to provide building blocks for growth, repair, or secretory products; and to furnish specific molecules, such as vitamins and some minerals, which are needed only in small amounts for certain metabolic processes but which cannot be synthesized by the body. Some of the vitamins are discussed in *Cell Structure and Function* by Loewy and Siekevitz (see Further Reading at the end of this chapter), in terms of their importance to the biochemical processes that occur within all cells. Although a wide variety of animal and plant material can be used to fuel the bodily machinery of animals, the great bulk of foods actually utilized falls into three classes: *carbohydrates*, *fats*, and *proteins*.

## *CARBOHYDRATES*

The molecule most widely used as a source of energy by living organisms is *glucose*, and its oxidation can be summarized by the chemical equation

$$C_6H_{12}O_6 + 6O_2 \rightarrow \rightarrow \rightarrow \rightarrow 6CO_2 + 6H_2O + \text{energy}$$

The series of small arrows indicates that many steps intervene between the initial stage, when only sugar, oxygen, and the appropriate enzyme or enzymes are present, and the final products shown at the right.

Most of the carbohydrates found in the bodies of animals are *polysaccharides* or polymers of simple sugars, usually glucose joined to another glucose, this in turn to a third, and so on until a huge molecule containing hundreds of glucose units is built up. Water is removed as each bond between successive glucose units is formed. The formation of a typical carbohydrate can thus be represented in its barest essentials by the reaction

$$n(C_6H_{12}O_6) \rightarrow \rightarrow \rightarrow \rightarrow (C_6H_{10}O_5)n + nH_2O$$

Among the most common polysaccharides are (1) the *glycogens*, which are formed in the cells of animals; (2) the *starches*, which are often formed in plant cells, particularly in those devoted to storage of food for the plant; and (3) *cellulose*, which is a much more stable and indigestable polymer than the other two and is formed by a different type of bond between successive glucose molecules (also involving the loss of water). Thus the formula $(C_6H_{10}O_5)_n$ serves only to identify carbohydrates as a class and not to distinguish among them; it describes equally well the main constituent of wheat flour (starch) or of paper (cellulose).

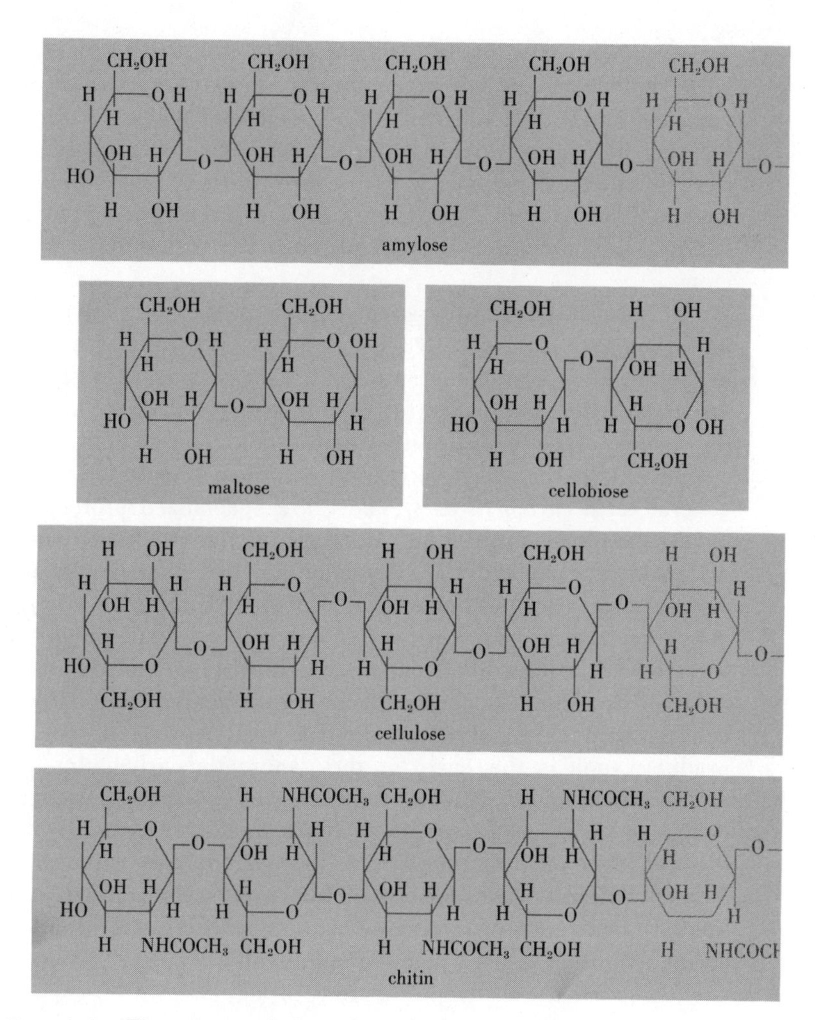


***Fig. 4-4*** *The structural formulas of several important carbohydrates and of the related compound, chitin. Maltose and cellobiose, dissacharides, are shown complete. The molecules of amylose, cellulose, and chitin continue on to the right, repeating the basic subunits an indefinite but large number of times.*

(See Fig. 4-4.) *Chitin*, the main stiffening constituent of insect exoskeletons discussed in Chapter 3, is a modified carbohydrate, and its oxidation is also an energy-yielding reaction. Hence certain animals are able to use chitin as a food. A curious use has been found for *inulin*, a polysaccharide found in the Jerusalem artichoke and related plants. In kidney function studies, inulin is injected into the mammalian bloodstream and cannot be metabolized but is filtered by the glomerulus but not acted upon by the renal tubule (see Chapter 8), thus giving a mechanism for measuring the glomerular filtration rate.

When most animals eat food containing polysaccharides, these large molecules are not taken directly into the cytoplasm from food vacuoles or through the cells of a specialized digestive tract. Instead they are first broken down either to simple sugars (such as glucose) or to small polymers containing two or three glucose units (or other simple sugars). The units of carbohydrate actually absorbed by animals from the digested food are usually simple sugars containing five or six carbon atoms.

Large carbohydrate polymers can be broken down to glucose units in a test tube, but only by such means as boiling with strong acids; in the digestive systems of animals, the process occurs at low temperatures and without the need for violent reagents. Digestive enzymes are the agents responsible for this gentler but more effective process—specifically, carbohydrate-digesting enzymes, sometimes called by the general term *carbohydrases*. These are specialized protein molecules having the important property of catalyzing the breaking up of the large molecule of carbohydrate to smaller polymers and finally to glucose molecules or units of similar size. Human saliva contains a type of carbohydrase that breaks down starch only part way and only by attacking one type of bond that joins together chains of sugar molecules; hence the result of its activity is a group of smaller polymers. Other carbohydrases can perform a complete breakdown with only glucose as the end product; such is the case for the pancreatic carbohydrases of most vertebrate animals. None of these carbohydrases, however, act on cellulose; excepting *Teredo*, the shipworm, and a few other wood-boring invertebrates, only microorganisms are currently known to produce *cellulase* to digest this widespread but very stable carbohydrate. The limited occurrence of cellulases has undoubtedly had broad influence on the evolution of both plants and animals.

***FATS*** Fats are compounds of *fatty acids* and *glycerol*. Each fatty acid consists of a chain of carbon atoms (frequently long or even branched) with hydrogen atoms attached and with an organic acid group at the end. In *saturated* fatty acids, each carbon has its full quota of hydrogen atoms. In *unsaturated* fatty acids, double bonds occur between one or more pairs of carbon atoms, meaning, of course, that each such carbon has one less hydrogen attached to it. Animal fats are more often saturated than vegetable fats. The degree of saturation affects the melting point. As a general rule, the less saturated fats remain liquid at lower temperatures. In the formation of a fat, three fatty acids combine with the three OH groups of glycerol in a kind of acid-base reaction, with the elimination of water; the resulting fat is an ester or organic acid "salt." The chain of the fatty acids commonly found in food is often about 16 carbon atoms long. This part of the fat molecule bulks so large compared to the glycerol part that the whole compound has properties somewhat like those of hydrocarbons such as petroleum or paraffin oils. The most important of these properties for long-chain fatty acids is insolubility in water. Short-chain fatty acids such as acetic acid are fully soluble in water. Fats are very compact fuel molecules for the use of living organisms because they yield more energy per unit weight than any other food. Carbohydrates, by contrast, are already partly oxidized. Thus fat, rather than carbohydrate, is often stored for future nourishment in the fat cells of animals and in the seeds of plants. Seeds are, of course, one of the major sources of edible fats for man.

The problem of digesting fats is not simply one of chemically breaking them down to fatty acids and glycerol; more important, it involves reduction in size of the droplets into which fats aggregate in an aqueous medium. The insoluble fats tend to aggregate into drops as the food is fragmented and other materials are removed in the stomach and intestine. Were fats to remain in this form, they would pass on through the digestive tract and be lost in the feces as, indeed, is the case in some diseases. For effective absorption from the intestinal lumen into the cells, these fat droplets must be reduced to the size range of colloidal particles such as the fat in milk—that is, to droplets one or a few microns in diameter. Fragmenting a 1-mm drop into $10^9$ droplets $1\mu$ in diameter increases by a millionfold the surface area of fat exposed to the surrounding water. Reactions that occur at the surface of the fat droplets permit the transport of fats across the intestinal wall.

*Bile salts* present in bile—which is the liquid that flows from the liver or from a storage depot, the gall bladder, into the small intestine of vertebrates—have the property of spreading out into thin layers between fats and water. A molecule of a bile salt is attracted both to the fat and to the water; that is, it has two parts, one tending to make it water soluble and the other tending it to make it fat-soluble. In the presence of both fat and water, these molecules slip into interfaces between the two and increase the area of such interfaces. This process is called *emulsification*. Such emulsified fats are exposed to the action of digestive enzymes and other chemical processing.

Fats in the intestinal contents are acted upon by enzymes manufactured in the pancreas, a large gland that lies close to the intestine and secretes pancreatic juice, which contains several important substances, into the intestinal lumen. The fat-digesting enzymes are called *pancreatic lipases*. These catalyze the breaking up of fats into the three fatty acids plus glycerol. Despite the availability of pancreatic lipases in the intestines of mammals, however, some of the emulsified fats penetrate through the layer of cells lining the intestine as complete fat molecules.

***PROTEINS*** Proteins are the third major foodstuff. These ubiquitous and almost infinitely varied macromolecules are a major feature of living protoplasm. As described by Loewy and Siekevitz (see Further Reading), these are polymers of amino acids, and the digestion of protein food consists of the breaking down of proteins first into smaller polymers and finally into the amino acid units themselves. Twenty-two common *alpha amino acids* make up most of the proteins found in animals and plants. (See Fig. 4-5.) These amino acids differ considerably in size and complexity, but since all of them have an organic acid or carboxyl group (COOH) proximate to an amino group ($NH_2$), all are exactly alike in one part of the molecule–that is, $CHNH_2COOH$. This key portion of the amino acid molecule is attached to other groups, which vary in size and chemical structure. For simplicity, the individually different parts of the several amino acids are customarily abbreviated as *R*; thus in this shorthand representation, all of the biologically important amino acids forming proteins are symbolized as $R$-$CHNH_2COOH$. When amino acids are joined together to form proteins, the carboxyl and amino groups on two different amino acids react to form what is called a *peptide bond* between the carbon atom of one and the nitrogen of the other. (See Fig. 4-6.) The product of this reaction, called a *dipeptide*, has many of the properties of a single amino acid; it can react in the same way with either the amino or the carboxyl group of a third amino acid to form a tripeptide. This process is, in fact, continued in the formation of proteins until hundreds of amino acids are linked by peptide bonds.

When proteins are consumed as food, they are broken down into their component amino acids before being taken through the walls of the digestive tract. In the digestive systems of highly organized animals, a number of specialized enzymes attack the bonds between certain amino acids but not those between others. Some enzymes may break long chains of proteins near the middle, others break off only terminal amino acids of a chain. Since some of the amino acids have additional carboxyl or amino groups within the entity *R*, peptide bonds can link these amino acids to others at more than one point, thus forming side chains. (See Fig. 4-7.) Some digestive enzymes work at or close to such branching points. As a result of this diversity of enzymes for protein digestion, protein molecules are split into a variety of fragments, which in turn are acted upon by other enzymes until the separation into single

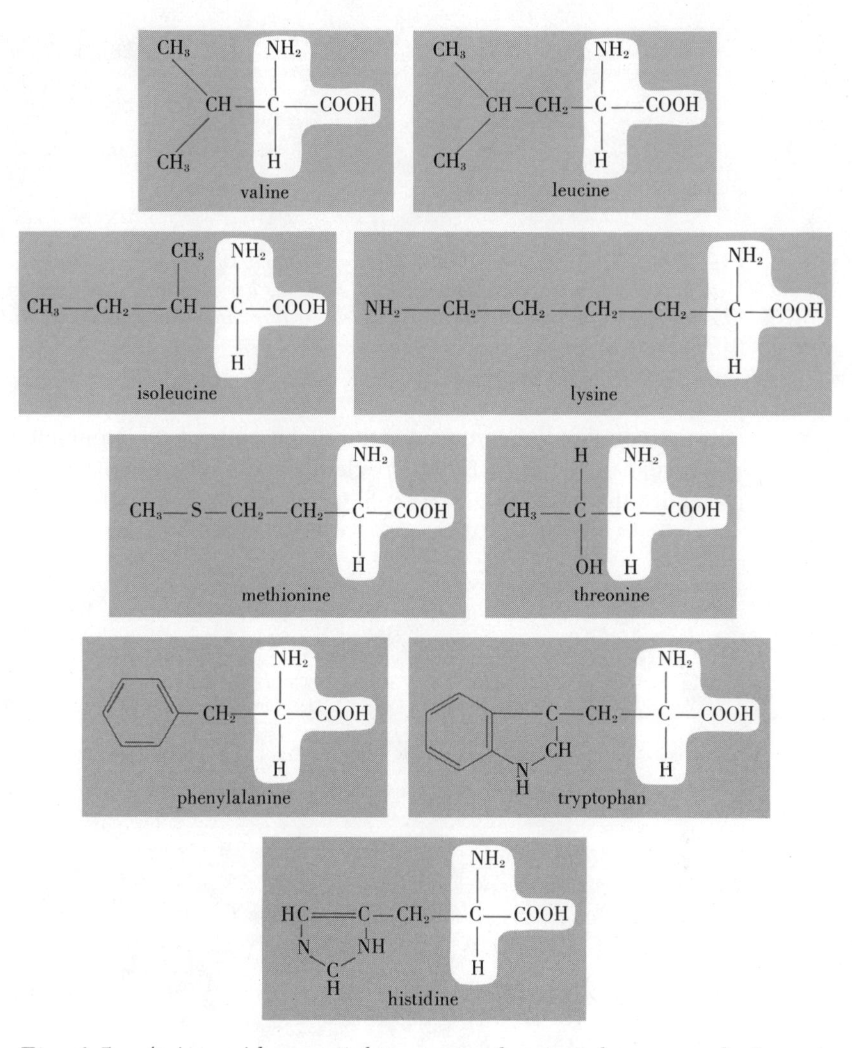


***Fig. 4-5*** *Amino acids essential to man and many other mammals. In each case the same active grouping appears with an amino group in the alpha position relative to a carboxyl group.*

amino acids is completed. In simpler animals, these steps all occur within food vacuoles or in single chambers, such as the gastrovascular cavity of coelenterates or the branched digestive tract of the flatworms. In more highly organized animals such as vertebrates or insects, however, different kinds of protein digestion occur at separate places along the "disassembly" line of the digestive system.

In vertebrates, the first steps in protein digestion occur in the stomach, where the protein molecules are broken into various kinds and sizes of *polypeptides*, usually in acid solution. The protein digestive

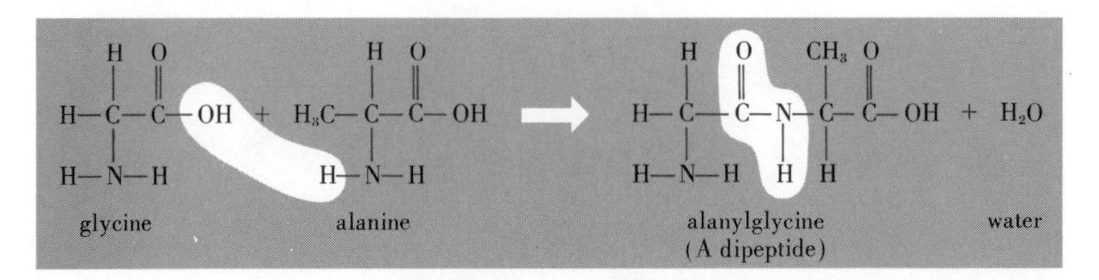


***Fig. 4-6*** *The reaction of two alpha amino acids to form a dipeptide. The reaction could also have occured in reverse, between the carboxyl group of alanine and the amino group of glycine to form the dipeptide, glycylalanine. (After Hoar.)*

enzyme commonly secreted by the gland cells of the stomach walls in mammals is called *pepsin.* Other glands secrete hydrochloric acid (HCl) to acidify the food in the stomach. Given enzymes usually function only

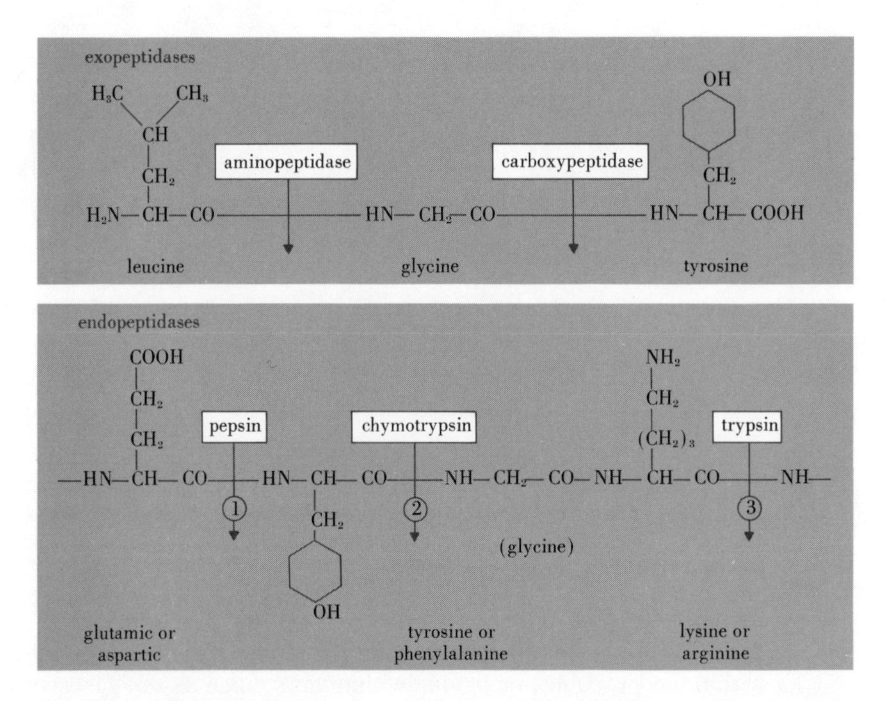


***Fig. 4-7*** *Exopeptidases (enzymes which catalyze the disassembly of protein molecules) act on peptide bonds adjacent to the ends of the molecular chain. The aminopeptidases act on the bond adjacent to a terminal amino group; carboxypeptidases on the bond adjacent to a terminal carboxyl group. The endopeptidases act on peptide bonds in the interior of protein molecules but only when the amino acids shown above participate in the bond. Chymotrypsin, for example, acts on an interior dipeptide bond only if tyrosine or phenylalanine participates in the bond. (After Hoar.)*

at an appropriate pH. Thus enzymatic activity can be started and also stopped by the secretion of acid or alkaline fluids. In the small intestine, conditions become slightly alkaline as a result of the secretions of intestinal wall glands. This stops the activity of the gastric enzymes. Now, other protein-digesting enzymes complete the breakdown of proteins and peptides that have escaped the action of the pepsin or have only been partly broken down. For the most part these enzymes are adapted to reducing peptides (short chains of amino acids) to single amino acids. Some of the intestinal protein-digesting enzymes come from glands in the wall of the intestine, but the most important single one, called *trypsin*, is produced in the pancreas. In mammals, the whole process of protein digestion is somewhat more rapid and efficient than in fishes, reptiles, or amphibians; one reason for this superiority is the greater division of labor among specialized protein-digesting enzymes. Chewing, body temperature, and greater efficiency of the circulatory and respiratory systems also have effects.

In general, large molecules such as proteins and polysaccharides are not absorbed whole from the gut. This is fortunate, for such molecules when introduced into the body by accident serve as *allergens*; that is, the body defenses treat these molecules as foreign materials and react against them. The severity of the reaction can be judged by the pain, fever, and swelling often caused by the injection, for example, of dead typhoid organisms (essentially protein molecules) intramuscularly for immunization purposes. These large molecules, furthermore, could normally not be used directly in any case, since they have been assembled for the use of another organisms. In addition, any system capable of absorbing or passing macromolecules would run the risk of leaking macromolecules as well. They must first be reduced to building block size and then resynthesized into the needed enzymes, structural proteins, and so forth.

## PROTECTION AGAINST SELF-DIGESTION

The efficiency of protein-digesting enzymes raises an important problem: what prevents pepsin and trypsin from digesting the cells that produce them or other cells that line the stomach and intestine? If these same cells were part of the food, they would be rapidly destroyed as their proteins were broken up into single amino acids. Individual cells, of course, die and are replaced, as in many other parts of the body, but the secretory systems remain intact for years, all the while producing enough pepsin, trypsin, and other enzymes to digest the whole animal many times over. This is an example of the complexity of chemical engineering as carried out by living organisms: manufacturing potent digestive enzymes is not enough; they must, in addition, be prevented from exercising their powers of protein breakdown until the appropriate time and place.

Pepsin and trypsin are neither synthesized nor stored as such in the gland cells of the stomach wall or the pancreas; instead, these cells secrete inactive molecules, *pepsinogen* and *trypsinogen*. When these reach the acid stomach or the alkaline contents of the small intestine, they lose a few amino acids on parts of the enzyme molecule; this loss *uncovers* the active surface of the enzyme so that it can go to work catalyzing the fragmentation of other proteins into their component amino acids. The presence of a little active pepsin or trypsin hastens this uncovering process. The *activation*, therefore, proceeds at a rapid rate once the pepsinogen or trypsinogen reaches the place where it is needed. Further protection is provided to the cells that line the digestive tract by a layer of secreted mucus that forms an effective barrier against the active digestive enzymes. The whole story, however, remains unknown and untold. A great variety of intestinal parasites, for example, live in these intestinal solutions containing activated digestive enzymes and thrive, nevertheless. Their protective mechanisms have not been uncovered.

## *THE ACTIVE NATURE OF INTESTINAL ABSORPTION*

So far little has been said concerning the way in which food molecules move into the body from the digestive cavity of the animal. Sometimes, after a heavy meal, there is a higher concentration of glucose, for example, in the intestine than in the blood that flows through the capillaries in the intestinal wall. In such cases, glucose molecules will diffuse from the region of high concentration to the place where they are less abundant. The end result of simple diffusion, however, would be an equal concentration of glucose in blood and intestinal contents. Even though glucose is rapidly removed from the blood by the liver and the cells of other organs, actual measurements have shown that there is usually a higher concentration of glucose in the blood than in the intestinal contents. That this sugar, nevertheless, moves fairly rapidly into the blood takes us by surprise, for the movement is against the *concentration gradient*. This is a clear example of "uphill" flow of a molecule. Other such examples of the flow of molecules against concentration gradients are also known and will be discussed in other chapters. Such movement of substances against a concentration gradient, one of the most important processes that distinguishes living from nonliving systems, is called *active transport*.

We should, of course, like to know how the active transport of glucose takes place, but as yet we can provide only partial explanations.

A basic principle of physical chemistry states that if a given substance is equally concentrated, initially, in two adjacent containers separated by a semipermeable membrane (here the cell membranes of the intestine and the blood vessels), work is required to separate the solution of the molecules of this substance into two portions, one of higher and the other of lower concentration. Active transport requires energy, and in living systems it is accomplished only by cells that are utilizing more food and have a higher rate of activity than they would were they not engaged in such active transport. Just how the energy of biochemical reactions is utilized to achieve active transport is not yet clearly understood, but the most widely held theories postulate some sort of carrier molecule that combines at one side of the cell with the glucose or other substance to be transported and is separated from it at the other side of the cell by local enzymes or by other (poorly understood) conditions there. Other models have been constructed but none have yet been shown to conform with living systems. It is quite possible that within the next few years much more will be learned about this universally important living process. Such knowledge will solve one of the most intriguing and fundamental problems of biology and ought, as well, to produce several Nobel laureates.

## *COMPARATIVE ASPECTS OF FEEDING*

Animals can easily (and superficially) be divided into *herbivores* and *carnivores*, although some are both and may be called omnivores (man, raccoons, bears, and cockroaches, for example). Among the herbivores, there are many specializations and these are of considerable interest because of the structural and functional differences that are likely to be identifiable in the different types. Herbivores may feed only on microorganisms such as bacteria, algae, or yeasts. Other forms are specialized for feeding upon leaves, bark, wood, fruit, seeds, flowers, pollen, or roots of higher plants. In some cases, only one species or genus of plant may be satisfactory. The koala bear, for example, a marsupial of Australia, feeds exclusively on eucalyptus leaves. Many insect larvae, particularly caterpillars, may be similarly restricted to a single food plant. Other animals—aphids and some mosquitoes, for example—feed exclusively on plant fluids. Some are parasitic and live part or almost all of their life in living plant tissue, often causing local tumors or foreign-body reactions called *galls*. In passing, note that the plants have often interacted with their animal predators in an exploitative fashion. Thus bees, moths, hummingbirds, and bats have all been utilized as cross-pollinating agents in exchange, so to speak, for being provided with nectar, pollen, and often nourishing petals. Fruit often serves to facilitate the distribution of seeds by animals. Plants have also reacted evolutionarily against their predators in a defensive fashion. The development of thorns or of siliceous particles in leaves, which wear down the teeth of grazers, or of offensive flavors or odors, or the production of toxic alkaloids or other poisons are common. Even the production of estrogens to interfere with the fertility of their predators has been alleged. Recently, many North American plants have been shown to produce substances that mimic the juvenile hormone of insects, thus, perhaps, preventing the maturation of larvae that feed upon these plants. Intriguing life histories—as of larvae that cease feeding on leaves and descend into the ground to feed on roots or detritus—may represent countersolutions to this defense.

Carnivores are at least as varied in action and reaction. Certain types of animals seem not only to be particularly abundant but to be particularly exploited as food by a large number of predators—earthworms, rodents, many insects, planktonic animals, and many mollusks, for example. But almost every animal has one or more fairly regular predators. Some such predators are themselves highly specialized, as in the case of many snakes that feed exclusively on birds' eggs. In *Dasypeltis*, an egg-eating snake from Africa, the specialized vertebral centra in the esophageal region are sharp ventrally and project into the esophagus where they cut open the egg shell as the egg is swallowed. The fluid portion is retained; the solids regurgitated. Some carnivores eat only individual parts of their prey. An extreme example is certain African cichlid fishes, which allegedly restrict their diet to fish scales; many animals suck blood or other body fluids.

Other animals cannot be classified either as herbivores or carnivores. Some, for example, feed on organic products—clothes moths on hair or fur, wax moths on beeswax. Many are *coprophagic*, feeding on the feces of other animals. Yet others feed on carrion or decaying plant material. Indeed, the hyena has been said to be particularly specialized for biting open carrion bones and feeding on the marrow inside. Finally, there are a large number of internal parasites adapted for living in many different environments but often in the bloodstream or in the intestinal tract of some other species. Some parasites may completely consume their host. This is especially common in insects. Curiously, many such internal parasites, always invertebrates, pass different parts of their life cycle in different hosts. In many cases the parasite is restricted to one particular species of host in each part of its life cycle. In other cases a wide variety of alternate hosts will be accepted.

Not uncommonly, animals with different life phases feed on different kinds of food in each phase. Sometimes a given phase does not feed at all but subsists entirely on previously stored food. Some adult insects do not even have mouth parts and are unable to feed. There are birds that normally feed on seeds as adults but catch insects for feeding their young during the period of rapid juvenile growth. Mammalian young are all milk-feeding, but guinea pig young may be weaned immediately, apparently without harm. Amphibian tadpoles are ordinarily herbivores, whereas the adults are usually carnivores. Such arrangements may not only represent a mechanism for reducing competition between the adults and their own young but may also or alternatively provide particular nutrients needed for growth or for reproduction. Perhaps, such variations in dietary habits may also represent the needs of an immature digestive system which still lacks fragmentation mechanisms, intestinal flora, or other key features.

Not surprisingly, animals have often evolved defense mechanisms against their predators and parasites. They are far too varied for us to do them justice here, in part because no one has had the time and ingenuity to recognize most of the potential or actual defense mechanisms. They fall, however, into behavioral, structural, and physiological types. Some animals actively avoid encountering their enemies by using specialized sense organs, by building protective walls or burrows, or by other barriers. Some digger wasps may even build false, accessory burrows, apparently to mislead other wasps that are nest parasites and thereby reduce the likelihood of predation by dissipating the parasites' time and energy. Some animals resort to camouflage. Camouflage among insects preyed upon by birds that hunt by vision is being intensively studied at present. One can only speculate, but profitably and intriguingly, on what the parallel defenses would be like in insects preyed upon by acoustically orienting bats. Here camouflage ought to be acoustic! Two examples of defense mechanisms in moths against bats are already known. In one family, the moths have evolved organs of hearing with which they detect the ultrasonic calls of hunting bats. When the bats are sufficiently close as to be threatening, these moths take up evasive flight patterns such as loops or power dives. In another family of moths, also provided with bat-detecting ears, a series of ultrasonic clicks are emitted upon hearing an approaching bat. These clicks appear to inhibit the bat's hunting success, but the mechanism has not yet been fully analyzed. Perhaps in the presence of such clicks, the bat cannot interpret the echoes it receives back from the insect. On the other hand, these moths have been shown to be distasteful to a large variety of insectivorous animals. Their clicks could be an acoustic version of warning "coloration." The warning would read: "Don't catch and kill me! I taste bad."

Protective covers (as in spit bugs or scale insects), armor (as in many catfish or armadillos), speed, bad flavor (in many insects, often associated with warning coloration), intelligence, defensive chemical discharges (in millipedes, stink bugs, roaches, and others), and high reproductive rates (as in many insects, bony fish, house mice, and rats) are examples of other ways in which predation has been compensated or reduced. A further analysis of predator-prey relationships and other aspects of feeding is given in *Ecology* in this series.

## FURTHER READING


Annison, E. F., and D. Lewis, *Metabolism in the Rumen.* London: Methuen, 1959.

Batra, S. W. T., and L. R. Batra, "The Fungus Gardens of Insects," *Scientific American*, 217 (5): 112–120, 1967.

Brower, L. P., "Ecological Chemistry," *Scientific American*, 220 (2): 22–29, 1969.

Brues, C. T., *Insect Dietary.* Cambridge, Mass.: Harvard University Press, 1946.

Davenport, H., *Physiology of the Digestive Tract.* Chicago: Year Book Medical Publishers, Inc., 1961.

Ehrlich, P. R., and P. H. Raven, "Butterflies and Plants." *Scientific American*, 216 (6): 104–113, 1967.

Hoar, W., *General and Comparative Physiology.* Englewood Cliffs, N. J.: Prentice-Hall, 1966.

Jennings, J. B., *Feeding, Digestion, and Assimilation in Animals.* New York: Pergamon, 1965.

Loewy, A. G., and P. Siekevitz, *Cell Structure and Function*, 2d ed. New York: Holt, Rinehart and Winston, 1969.

Morton, J., *Guts. The Form and Function of the Digestive System.* New York: St. Martin's, 1967.

Pike, R. L., and M. L. Brown, *Nutrition: An Integrated Approach.* New York: Wiley, 1967.

Romer, A. S., *The Vertebrate Body*, 3d ed. Philadelphia: Saunders, 1962.

Rothschild, M., and T. Clay, *Flees, Flukes, and Cuckoos.* London: Collins, 1952.

Solomon, A. K., "Pumps in the Living Cell," *Scientific American*, 207 (2): 100–108, 1962.

Wilson, T. H., *Intestinal Absorption.* Philadelphia: Saunders, 1962.

chapter 5

# Metabolism and Respiration

Living organisms carry out their intricate and balanced activities by the controlled release of energy from the oxidation of food stuffs. The general term for these oxidative processes is *metabolism*. Metabolic rates are expressed in units of energy released per unit time (equivalent to power)—for example, kilocalories per hour. The processes of bringing oxygen to the cells of an animal and carrying away carbon dioxide ($CO_2$) are quite as essential as the digestion and absorption of food. Water, like carbon dioxide, is also a product of metabolism; under some conditions, water formed by the oxidation of foods supplies the entire needs of animals that live in extremely dry situations. Some small desert mammals—such as the kangaroo rat, for example—live for months without drinking water, eating only dry food, such as seeds. Although they lose water steadily in their urine and feces and by evaporation from the skin and from the lungs, this loss is balanced by the so-called metabolic water derived from the oxidation of the hydrogen atoms in their food. Most terrestrial animals must obtain additional water by drinking or by absorption through their skin. For aquatic forms, obviously, the challenges are quite different and the solutions are often ingenious. (See Chapter 8.)

The exchange of oxygen and carbon dioxide between active cells and the outside environment is usually called *respiration* or *respiratory gas exchange*, since both oxygen and carbon dioxide are gases at ordinary temperatures. The organs for their exchange with the environment form respiratory systems. The term *respiration* is also used for the biochemical events that occur inside living cells in the stepwise breakdown of food molecules and the transfer of resulting energy to other molecules. This important activity is best called *cellular respiration*. Food molecules can be stored within the body of an animal, but only in special and limited cases (such as in diving animals) is oxygen stored very briefly against future need or carbon dioxide allowed to accumulate for future disposal. Respiratory gas exchange is a crucial and continuous process. The efficiency of this process and the resistance of the respiratory system to failure under adverse conditions determine, in large measure, the success and survival of the animal in question. The rate of exchange, of course, need not be the same in all animals but depends on the total metabolic activity of the organism. Such activity not only varies from species to species according to the way of life of the animal but also varies from time to time in a given individual depending on its reproductive stage, its locomotor activity, etc. Some examples of metabolic rates in organisms of markedly different ways of life are shown below.

Oxygen consumption of various animals (in milliliters of $O_2$/gram/hour)

| Animal | $O_2$ | Animal | $O_2$ |
|---|---|---|---|
| *Paramecium* | 1 | *Calliphora* (insect, blowfly) | 1.7 |
| *Amoeba* | 0.2 | *Vanessa* (insect, butterfly) | 0.6 |
| *Aurelia* (coelenterate) | 0.0034 | *Asterias* (echinoderm, starfish) | 0.03 |
| *Ascaris* (roundworm) | 0.5 | Goldfish | 0.07 |
| *Octopus* (mollusk) | 0.09 | Trout | 0.22 |
| *Arenicola* (annelid) | 0.03 | Rat | 0.95 |
| *Uca* (crustacean, crab) | 0.05 | Cat | 0.44 |
| *Homarus* (crustacean, lobster) | 0.5 | Man | 0.2 |

Such a table has less utility than would appear at first glance. One of the basic problems of such a comparison of metabolic rates is that the structure of these organisms is so different. Some have shells or other inert or relatively inert weight. Some have high water or fat contents. To some extent, such structural differences can be compensated for by expressing oxygen consumption in terms of fat-free, dry weight or grams of protein. We are not told, however, of the conditions under which oxygen consumption was measured—such as temperature, time of day, tide, light or darkness, season, whether the animals were fed or fasting, their state of nutrition, the phase of their life cycle, and whether quiescence on the part of the subjects was achieved during the measurements. We do not mean to belittle such comparative data but only to warn of some of their limitations.

Although oxygen and carbon dioxide are commonly thought of as gases, their primary biological roles are played in aqueous solution. Oxygen remains as $O_2$ whether it is in the gas phase or in solution. Only a relatively small amount will dissolve in water at normal temperatures and pressures. When air, which contains just under 21 percent of oxygen, is in contact with water at near freezing temperatures, approximately 1 milliliter (ml) of gaseous oxygen will dissolve in 100 ml of water. Only about half as much is held in solution at the body temperature of a bird or a mammal. Carbon dioxide, on the other hand, reacts with water chemically to form carbonic acid, $H_2CO_3$, which in turn dissociates rapidly to form $H^+$ and $HCO_3^-$ (bicarbonate) ions. In water and in tissue fluids that are neither strongly acid nor alkaline, most of the carbon dioxide exists, therefore, as bicarbonate ions rather than as dissolved $CO_2$ molecules. Because water reacts in this way with $CO_2$, it can hold much more carbon dioxide (as bicarbonate ions) than it can hold oxygen in solution.

## *RESPIRATORY EXCHANGE BY DIFFUSION AND CIRCULATION*

Unicellular animals and others of small size have no special organs for respiration. The use of oxygen or the production of carbon dioxide by reactions occurring within a cell sets up concentration gradients: oxygen becomes less concentrated inside the cell (and at particular sites within a cell) and carbon dioxide more so as a result of the oxidation of food molecules. Oxygen and carbon dioxide move along such concentration gradients by diffusion. This simple physical process usually suffices for gas exchange in cells which are within a fraction of a millimeter of surrounding oxygenated water. Cells cannot operate at too low a concentration of oxygen, and excessive amounts of carbon dioxide are injurious. In addition, diffusion is a relatively slow process and were the distances to be covered to be considerable, diffusion could not suffice. The actual rate of diffusion of oxygen in vertebrate connective tissue is about 0.0001 ml/cm$^2$/cm/atm. Oxygen diffuses some 10,000 times faster in air. Hence a reliance on diffusion as a means of transporting the respiratory gases sets a limit of roughly 1 mm to the thickness that can be attained by an unspecialized animal. The difficulty can be alleviated by lowering the metabolic rate, and hence the rate at which oxygen is consumed, but this in turn severely limits the activity of the animal and its ability to obtain food or escape from enemies. This limitation on size applies not to total volume but only to the thickness or separation of the cells from the medium—that is, the distance from the outside water to the deepest parts of the organism that are carrying on active metabolism. Therefore flat or sac-shaped organisms such as flatworms and coelenterates may be quite large. Inert skeletal material or the jellylike tissues that lie between the inner and outer cell layers of coelenterates may be farther removed from the surface than can muscle cells or nervous tissue, since the inert material does not metabolize.

Another important fact about diffusion as a means of respiratory exchange in small animals is that it suffices regardless of the complexity of the animal or the phylum to which it belongs, provided only that the animal is small and its surface permeable to oxygen and carbon dioxide. Thus early stages in the development of complex animals, including ourselves (before the embryo implants in the uterine lining), require no specialized respiratory organs. Most aquatic adult members of the phylum Arthropoda possess specialized gills and associated blood vessels to obtain oxygen from their environment, but very small arthropods, even though fully adult, often lack such respiratory structures altogether. All their respiratory exchange can easily take place by diffusion through the skin.

The size limit set by diffusion of oxygen and carbon dioxide can be greatly extended if the protoplasm or other, noncellular fluids such as blood are in motion. Over distances of more than about 1 mm, respiratory gases and other molecules can be stirred about and thus supplied where needed much more readily by circulation of fluid than by diffusion. In simpler, multicellular animals, such as coelenterates, the water within the digestive cavity or other internal cavities serves this purpose. The branched digestive cavity of a flatworm and the body cavities of roundworms and annelid worms contain fluid that is circulated in a rather irregular fashion when the muscles of the body wall contract. Such internal circulation aids in respiratory gas transport, but a great improvement in efficiency occurs when it is supplemented, as it is in most of the annelid worms and in the phyla Mollusca, Arthropoda, and Chordata, by an organized circulatory system in which blood is pumped through more or less definite channels. Such circulatory systems will be considered in Chapter 7, but the functioning of respiratory and circulatory systems is so intimately interwoven that each of these systems must be studied with its interrelationships firmly in mind.

## GILLS

A skin that allows free diffusion of oxygen and carbon dioxide cannot fail to be rather permeable to other small molecules as well. This poses less critical problems for marine animals, in which the inorganic constituents and osmotic pressure of the body fluids are not greatly different from sea water, than for animals that live in fresh water, in the soil, or on land. These must maintain their internal fluids in a very different state from the surrounding environment. The better the protection afforded by a specialized skin, the poorer the skin necessarily becomes as a surface for respiratory exchange. For example, no animal has developed a skin that permits free exchange of oxygen and carbon dioxide while remaining impermeable to water molecules. It is therefore not surprising to find that the more active types of animals, and those equipped with protective skins, have specialized portions of the body's surface devoted specifically to respiratory exchange.

The simplest of these specialized surfaces are merely areas where the external surface is folded or formed into projecting appendages covered by a thin, permeable skin under which are located thin-walled blood vessels or spaces containing a fluid that can carry oxygen and carbon dioxide as it circulates within the animal's body. The folded or redundant surface of *gills* greatly increases the surface area through which exchange may occur. An interesting example can be found in certain annelid worms, which secrete or consolidate around their bodies a hard, protective tube. Such tubes are impermeable to respiratory gases. Many of these worms have fine, threadlike gill filaments that project outside of the tube from time to time to allow respiratory gas exchange between the blood that flows through them and the external water. Others have shorter appendages that serve as gills. Filamentous gills are especially prominent among some of the marine worms that live in muddy layers at the bottom of the ocean, where oxygen is often very scarce or totally absent because of bacterial activity. The gills extend above the mud to oxygenated water. In other cases, such gills occur in worms that live in burrows in the mud or sand and do not secrete shells around their bodies. In all cases, the gills are retracted into the tube or burrow when danger threatens or when the water above becomes strongly agitated, or is absent, as at low tide.

Among vertebrates, some salamanders, many frog and toad tadpoles, and some fish also have external gills. These forms often also respire via the moist skin, the mouth and throat lining, and even the cloaca. Combined with effective circulatory systems, these sites suffice for most of the respiratory gas exchange required by a relatively quiet life. External gills, in order to expose effectively a large surface area, must often float loosely in sheltered water and, hence, are vulnerable to injury in an actively moving animal. They are also exposed to predators and parasites. In their place, most of the larger and more active types of aquatic animals have their gills located inside a cavity. These cavities always have openings to the outside water and often have means for circulating fresh water over the gills in order to facilitate respiratory exchange. The three major groups of highly organized aquatic animals possessing such protected gills are the mollusks, the crustaceans (a class of arthropods), and the fishes. Although all three have gills that accomplish the same function, the particular arrangements are quite different.

In most of the mollusks, the gills lie inside a *mantle cavity*, into which also open the mouth and anus. (See Fig. 2-4.) In the clams (whose gills are used primarily for feeding) and snails, this cavity is inside the hard shell; water is circulated through it mainly by the coordinated beating of millions of cilia. These motile organelles are similar to the cilia or flagella of protozoans, but in mollusks they are parts of specialized cells covering the surface of the gills and/or lining other parts of the mantle cavity. In the more active squid and octopus, the mantle cavity has thicker and more muscular walls, which actively pump water in and out and thus ventilate the gills. In the squid and octopus there is also a circulatory system, in which blood is pumped through the gills to carry oxygen and carbon dioxide to and from the rest of the body.

In crayfish, lobsters, crabs, and many smaller crustaceans, the gills are relatively large and lie on the two sides of the thorax or middle portion of the body, lateral to the body proper, but inside the main plates of the chitinous exoskeleton. (See Fig. 5-1.) As in the mollusks, the circulatory system pumps blood through the gills. Some crustaceans have other types of gills derived from portions of their jointed appendages, but the type illustrated diagrammatically in Fig. 5-1 is typical of the larger and more active members of this important group of animals.

Among the vertebrates, aquatic respiratory systems are most highly developed in the fishes. The gills are located on the edges of a series of lateral openings from the *pharynx*, that part of the digestive system just posterior to the mouth. There are often four or five of these openings or *gill slits* on each side of the pharynx. Anteriorly, there may also be a pair of modified slits, the *spiracles*, one on each side, often specialized for water intake in bottom dwelling fish such as the rays. In the sharks and rays, the gill slits open directly to the outside; in most of the bony fish, however, the several gill slits open into an

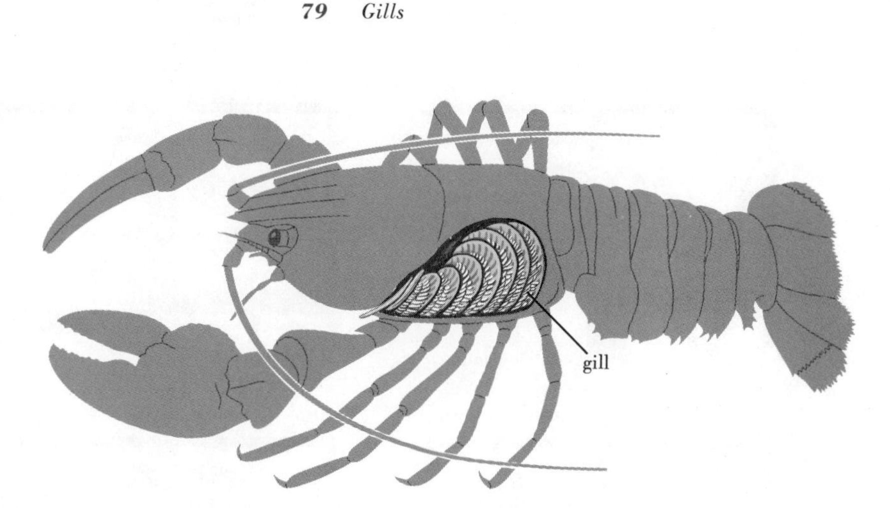


***Fig. 5-1*** *The gills of a lobster (class Crustacea). The exoskeletion has been cut away to show the gills in relation to the body.*

external chamber on each side, lateral to the pharynx. These lateral chambers are really outside the body of the fish, but are covered by or partly enclosed within a bony plate, called the *operculum*. A restricted opening at the posterior edge of the operculum is the obvious external feature of the gill system of most familiar fish. Between adjacent gill slits on each side is located a skeletal *gill arch*, by which the *gill filaments* are supported and through which they are supplied with blood vessels.

Water is circulated from the mouth through the gill slits and out through the *opercular cavity* of bony fishes by two mechanisms. The opening and closing movements of the fish's mouth bring water in; most of the water is then forced out over the gills by muscular contraction of the buccal and pharyngeal walls. Because soft flaps of connective tissue around the mouth opening act as check valves, preventing water from flowing back, out of the mouth, the water must pass over the gills. In a coordinated fashion, the operculum is moved outward by muscles attached to its outer surface. This enlarges the opercular cavity and also draws water past the gills. While the mouth is being opened, another valvelike flap of tissue at the posterior edge of the operculum prevents water from flowing back into the opercular cavity. (See Fig. 5-2.) By either or both of these types of pumping action, most fishes circulate water through the mouth and over the gills. In some very active swimmers, such as mackerel, the movement of the fish through the water causes most of the current past the gills. Such active fish may not stay still for long periods of time because, in the absence of such ventilation of the gills, they fail to obtain enough oxygen.

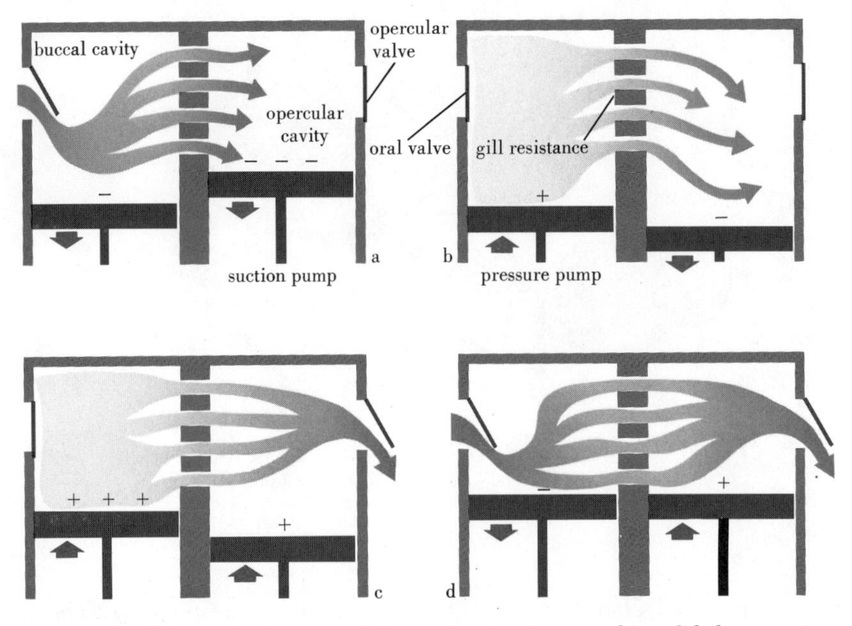


***Fig. 5-2*** *Schematic drawing of the mechanism in many bony fish for pumping water over the gills. The buccal (or mouth) chamber and pharynx are represented by the cylinder at the left of each pair; the opercular chamber is represented by the cylinder at the right. The movable piston of the bucco-pharyngeal chamber represents the musculature of the floor of the mouth and of the pharynx. The movable piston of the opercular chamber represents the muscles which regulate the volume of the opercular chamber. The opening at the left represents the mouth (guarded by a valve); the opening at the right is the opercular aperture. The wall between the two cylinders is perforated by the gill slits.* (Upper left) *Oxygen-rich water is drawn into the mouth by expanding the bucco-pharyngeal cavity while the opercular aperture is closed.* (Upper right) *The water is pushed through the gill slits by contraction of the bucco-pharyngeal walls while the opercular cavity expands. Passing through the gills, the water yields part of its oxygen load to the blood.* (Lower left) *The valves of the mouth prevent egress as the opercular cavity is contracted. The water therefore passes out through the opercular aperture.* (Lower right) *The bucco-pharyngeal cavity is again expanded, thus drawing water in through the mouth. (After G. M. Hughes,* Vertebrate Respiration, *Harvard University Press, 1965.)*

Not all fishes, however, always live in water with an oxygen content adequate to support life. Hot swamps in particular tend to be oxygen poor. Oxygen grows less soluble as the temperature of the water rises, oxygen is rapidly depleted by the decay of plant and animal debris, and the water is often sufficiently sluggish as to offer only minimal circulatory aid to the diffusion of oxygen. Furthermore, such waters and related, quiet, freshwater ponds or streams may be opaque or partially so because of suspended organic material or suspended soil particles. Such opacity precludes or reduces photosynthesis in aquatic plants and further reduces the likelihood of there being adequate dissolved oxygen to support fish life. Thus, freshwater fishes have often evolved special respiratory systems where gills, no matter how efficient, could not serve the purpose of taking up oxygen. All of the known solutions—other than to assume a sluggish way of life, to *estivate* (a state of suspended activity resembling hibernation but associated with hot and/or dry seasons), or to exploit the thin film of water just under the surface (widely used by hatchlings)—involve air breathing. The fish usually rises to the surface, gulps a bubble of air, and passes this into specialized vascular chambers for absorption of the oxygen (via a film of water on the tissue surface). Such chambers are usually part of the gut or outpocketings of the gut and examples are known ranging from the mouth, pharynx, and stomach to the intestine and even the cloaca. The *lungs* of lung fish are also such structures. In many species of air-breathing fish, the gills may also function but are often reduced in extent. Under some conditions—extreme oxygen depletion of the water, for example—the gills would only serve to surrender oxygen from the body to the water as it was absorbed from the air bubble. In such cases, the gills are probably bypassed physiologically by reducing or cutting off their blood supply.

Conversely, some vertebrates occupy niches with unusually high concentrations of dissolved oxygen, such as rapidly flowing, cold, and turbulent mountain streams. Here respiration may be so "easy," especially for a sluggish way of life, as to have allowed the elimination of gills and lungs altogether (in some salamanders) and to shift respiratory function entirely to the skin.

## *COUNTERCURRENT EXCHANGE*

In the gills of fishes, and also in those of some mollusks, blood is carried close to the gill surface in the opposite direction from that in which the water is flowing. This *countercurrent* permits a more efficient exchange of oxygen between blood and water. As the blood enters the gill, it has a very low oxygen content. As the water enters the gill region, it has a relatively high oxygen content. This, of course, is why diffusion of $O_2$ from water to blood occurs. Were the water and blood to flow in the same direction, side by side, barely separated by gill and capillary walls, the maximum exchange that could occur, if there were time to reach equilibrium, would be for the concentration of oxygen finally to be the same in the blood and the water as both left the gill region. With a countercurrent system, much greater exchange is accomplished. In actuality, the blood enters at the opposite end from the water and flows in the opposite direction. Thus, blood that has already taken up some oxygen is exposed to the freshest water containing the highest concentration of dissolved oxygen just before leaving the exposed surface of the gill. At the same time, water that has already lost much of its oxygen is exposed to blood with practically no oxygen at all just before the water leaves the region of the gill. That is, entering water exchanges with exiting blood and entering blood exchanges with exiting water. The result of such an arrangement can be the almost complete transfer of oxygen from the water to the blood instead of a 50-50 distribution. (See Fig. 5-3.)

Similar countercurrent systems are known to occur widely among animals for the efficient exchange of a variety of blood-borne substances and heat. Heat, for example, is conserved in many mammals and birds by having the arterial blood running to the exposed limbs (deep in snow or immersed in cold water, perhaps) pass close to the veins returning from the limbs. The arterial blood then yields heat to the venous blood that is re-entering the trunk, and heat is thus retained in the core of the body. The limbs, of course, will be cool. The problem of supporting metabolism in such cool limbs has to be solved separately; the solutions are not yet well understood. Countercurrent exchange has also been hypothesized as functioning physiologically in gas secretion in the swim bladders of deep-sea fish and is used in the production of hyperosmotic urine in mammals. (See Chapter 8.) Most likely such a valuable and practical system has been far more widely invoked than is yet known.

## RESPIRATORY SURFACES FOR AIR BREATHING

Air contains about 21 percent oxygen, in contrast with the small amount that water can hold in solution—between 0.5 and 1 volume of gaseous oxygen per 100 volumes of water that has come into equilibrium with the atmosphere. Air is also easier to pump. Hence it would seem at first thought that animals living in air should have relatively simple respiratory problems. This is not the case, however,

***Fig. 5-3*** *A portion of a gill of a bony fish. The gill filaments support large numbers of capillaries which are intimately exposed in thin lamellae to water flowing by. The water flows counter to the direction of flow of the blood in the capillaries, optimizing the exchange of oxygen (see the inset above right). If water and blood flowed side by side in the same direction and if oxygen is exchanged by passive diffusion, the concentration of oxygen in the blood leaving the gills could not exceed that of the water leaving the gill region. By providing opposing currents, much higher blood oxygen levels can be reached. At each point the water has a higher concentration than does the blood flowing by. Thus, oxygen diffuses into the blood throughout the region of contact. The numbers shown are simply examples of relative oxygen concentrations which might occur in the two compartments under the two different conditions of parallel flow. The actual concentrations achieved depend on the speed of flow and the length of contact of water and blood.*

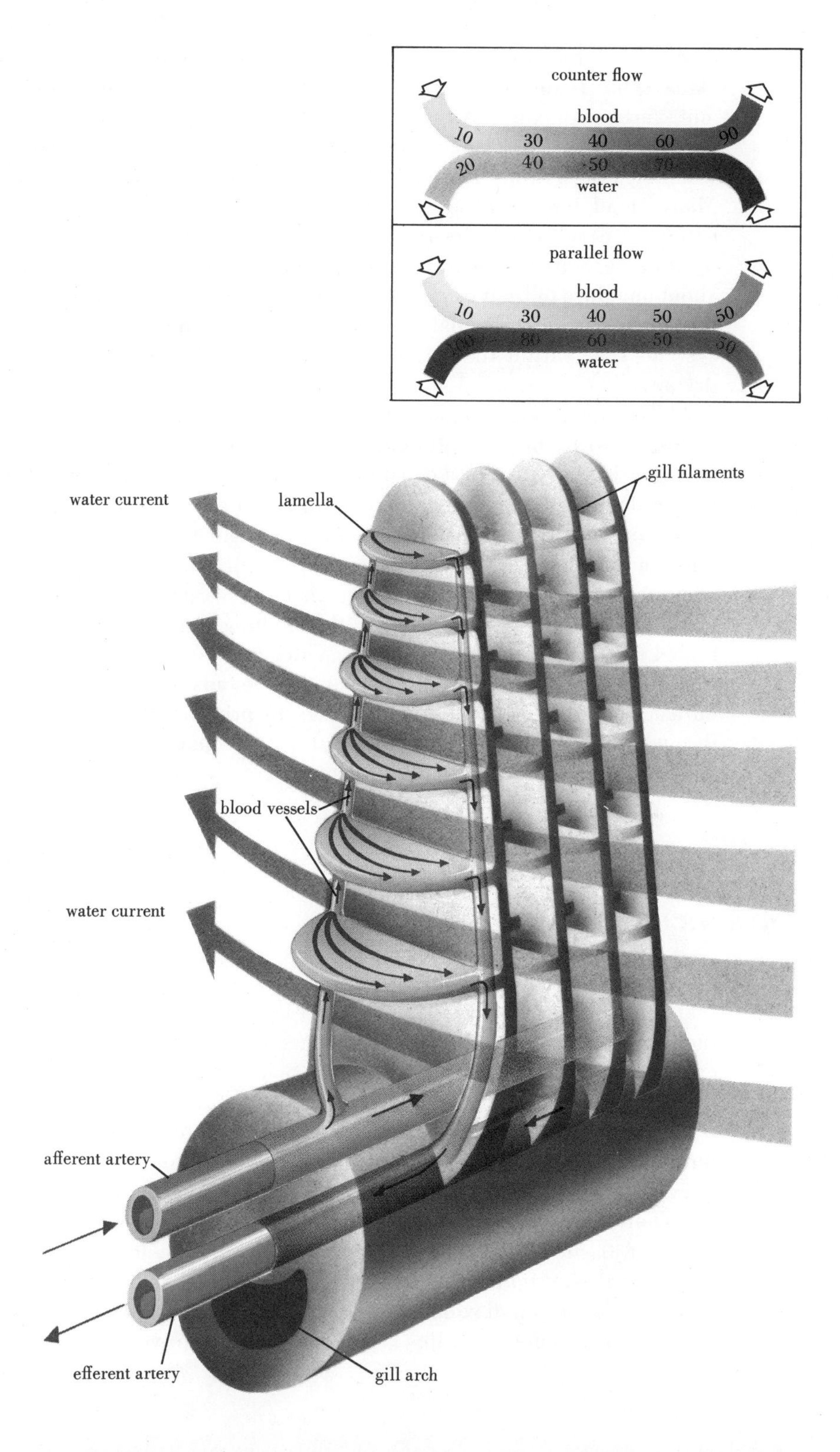
counter flow
blood
10
30
40
60
90
20
40
50
70
water
parallel flow
blood
10
30
40
50
50
100
80
60
50
50
water
gill filaments
water current
lamella
blood vessels
water current
afferent artery
efferent artery
gill arch

since terrestrial animals, to avoid drying out, require less permeable skins. This requirement necessarily reduces the diffusion of oxygen and carbon dioxide. Any living surface through which respiratory exchange can take place is also subject to the danger of water loss, whether it be in a human lung or on the moist body surface of a worm. Many small terrestrial animals do have moist skins and survive for extended periods only by remaining in places where the air is nearly or quite saturated with water vapor. Some frogs, for example, take up about one third of their oxygen through their skin. But many vertebrates, the terrestrial arthropods, and some of the snails have respiratory systems that permit them to spend their whole lives in either moist or dry air.

The terrestrial snails are good examples of animals that are equipped with small mantle cavities, which are called *diffusion lungs* because they are kept at a constant volume with very little active pumping of air in and out. The opening is rather restricted in size, and part of the inner surface has blood vessels close to the air. Since oxygen diffuses thousands of times more rapidly through air than through water or protoplasm, only rather large animals need to circulate the air within a hollow respiratory organ. Many diffusion lungs have the inner surface folded to increase the surface area. Water loss is reduced by the small size of the opening of the diffusion lung to the outside air, and the rapid diffusion of oxygen through air suffices to bring in all that is needed through the small opening. Among the Arthropoda, scorpions, some spiders, and isopods also have diffusion lungs.

## TRACHEAL RESPIRATORY SYSTEMS

Insects and a few other arthropods (centipedes, millipedes, and some spiders, for example) have a highly specialized type of respiratory system that differs significantly from the diffusion lung in that the oxygen is delivered directly to the cells and is not carried in the blood. In several of the body segments, insects have paired openings called *spiracles*, one on each side of the body. These communicate with air-filled cavities from which extend small branching tubes called *tracheae*. The branching is repeated over and over again until the terminal tubes, called *tracheoles*, are often less than 1 $\mu$ in diameter. Except at their very ends, the tracheae and tracheoles are filled with air. They ramify to every part of the insect's body. The more active an organ or tissue, the more abundant are the tracheoles supplying it. In some cases, tracheoles even penetrate individual cells. (See Fig. 5-4.) Oxygen and carbon dioxide travel along the tracheal tubes by diffusion. The larger air-filled cavities, however, are often ventilated by the

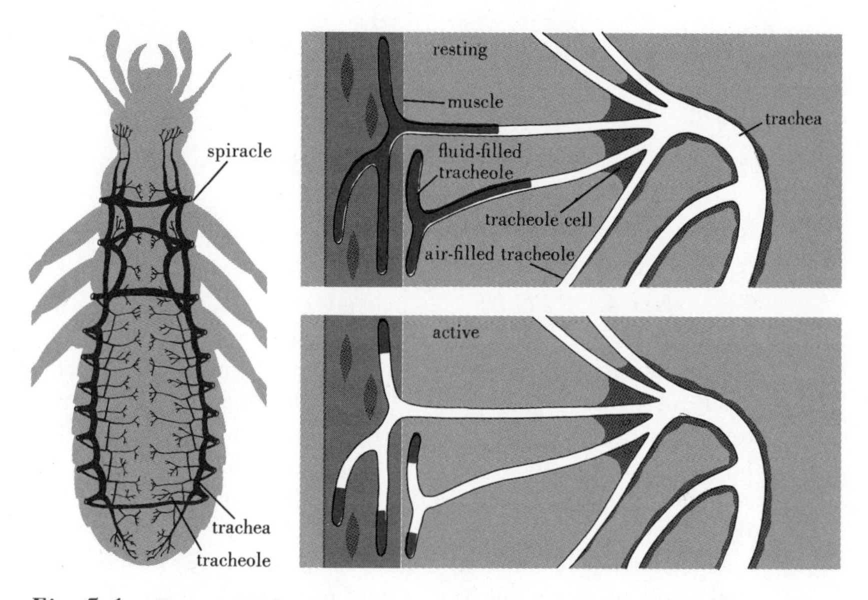


***Fig. 5-4*** *The tracheal respiratory system of insects.* (Left) *A greatly simplified view of the tracheal system. Air enters a series of spiracles on each side by diffusion. The tracheae and tracheoles branching from the spiracles allow the diffusion of air close to all parts of the body. The tracheal system is completely lined with cuticle which is molted along with the rest of the exoskeleton.* (Right) *Tracheolar endings are shown in a resting and in an active insect. (After Wigglesworth.) The terminal portions in both cases are fluid filled. Thus the oxygen first dissolves and then diffuses into the insects's body proper. During activity, however, with increased ventilation, the air penetrates further into the tracheoles and exchange is hastened. (Figure at left after P. Meglitsch,* Invertebrate Zoology, *Oxford University Press, 1967.)*

movements of surrounding muscles or exoskeleton. The spiracles are closed and opened by valves that are in turn operated by minute muscles. Each spiracle also has hairs around its edges that prevent the entry of dust particles and parasites.

This tracheal respiratory system of insects provides a direct avenue for respiratory exchange between the outside air and the cell, where metabolic reactions take place. It is an efficient system and, in insects, blood need not carry oxygen or carbon dioxide. A tracheal respiratory system would probably not suffice for the needs of a large animal unless there were extensive ventilation, for even in air there are limits to the distance over which diffusion can carry enough oxygen along minute tubes to supply high metabolic needs. A flying insect, in particular, uses oxygen at a much higher rate per gram of body weight than any bird. In these insects, the tracheal system of the flight muscles is automatically ventilated by the rhythmic compression of air sacs in the musculature. The tracheal tubes are lined by cuticle which must be separated and shed at each molt along with the rest of the exoskeleton. The need for such periodic molts and the limitations set by diffusion along such fine air-filled tubules may set limits on the ultimate size that insects can attain.

## THE LUNGS OF FISHES, AMPHIBIANS, AND REPTILES

During the first hundred million years or more of its history, our own subphylum was composed exclusively of fish. Only after vertebrates had become a highly successful aquatic group did terrestrial, air-breathing forms appear. The ancestral group of Crossopterygian fish from which the amphibians and all the tetrapods evolved were apparently tropical, swamp-dwelling forms with lungs much like those of present-day lung fish. Whatever the actual selection pressures that molded the evolutionary future of this group, the ultimate adaption was to terrestrial or partially terrestrial life—a change of way of life that must have opened many new feeding niches, new ways of protecting young, and other such previously unexploited improvements. Though our contemporary lung fish do not emerge onto land, African lung fish can survive long arid periods when the streams in which they live dry up by digging into the muddy bottom, building a kind of cocoon of mud and mucus, and entering a state of low metabolism called estivation. They remain in this condition until the flow of water recommences. During this period they are entirely dependent on pulmonary absorption of atmospheric oxygen.

Adult amphibians often have paired lungs connected to the pharynx through a muscular valvelike chamber, the *larynx*. The larynx is arranged so as to prevent water or food from entering the lungs. A single tube, the *trachea*, leads from the larynx and branches into two *bronchi*, each leading to a lung. Amphibian lungs are fairly simple sacs with only a moderate folding of the inner surface to increase the area for exchange of oxygen and carbon dioxide with the rich supply of small blood vessels coursing through the lung lining. The lungs do not contain muscles to pump air in and out; pumping is accomplished by contractions of the muscles of the body wall or by modified swallowing movements that force air into the lungs. The lungs are, however, elastic. When muscular contractions cease, the lung tissue tends to restore itself to its previous position, thus forcing air back out.

Not all amphibians, however, possess lungs nor do all of those with lungs use them regularly. Larval amphibians, which are almost always aquatic, depend on external gills and dermal respiration. Adult aquatic amphibians may retain external gills. Some of these never develop lungs at all, whereas others have both systems. All terrestrial adult amphibians depend substantially on dermal, buccal, and pharyngeal gas exchange and often have interesting arrangements of their circulatory system to facilitate such respiration. Carbon dioxide, which moves much more quickly across living membranes than oxygen, probably facilitated by enzymatic action, appears to be handled almost exclusively by extrapulmonary respiration in this group.

All reptiles have lungs; none are known to have either internal or external gills. In aquatic reptiles such as the soft-shelled turtles, however, dermal, pharyngeal, and cloacal respiration are used when the animal is submerged. Reptilian lungs, like those of amphibians, are rather simple sacs, only moderately convoluted and partitioned internally. The maximum respiratory exchange in reptiles seems to be rather low, presumably limiting vigorous activity to short bursts.

***MAMMALIAN LUNGS*** In mammals, the lungs are larger and so highly honeycombed with branching air passages and folded walls that there are no large, uninterrupted volumes of free air. Instead the air is confined to small chambers known as *alveoli*, roughly 1 mm in diameter, each located at the end of a terminal branch of one of the bronchi. That is, in the mammalian lung, the bronchi divide and redivide and divide yet again until they form microscopic passages called *bronchioles*, each leading to an alveolus. (See Fig. 5-5.) In a frog, there are no more than 20 sq cm of lung surface

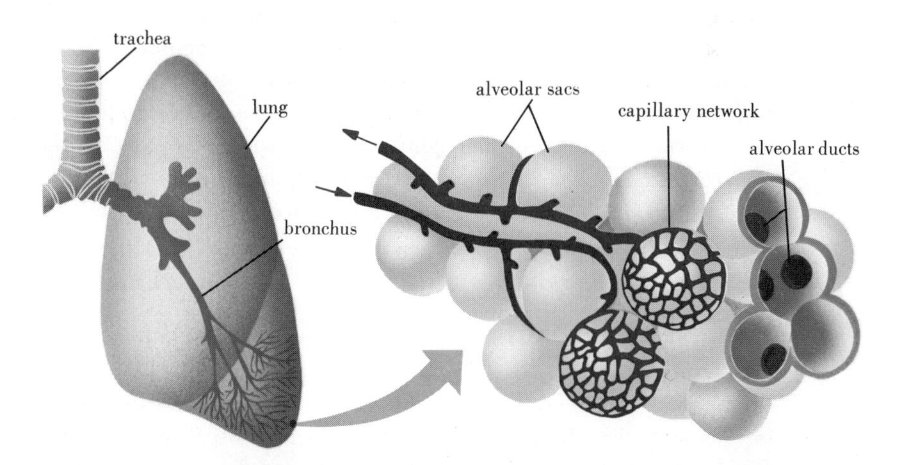


***Fig. 5-5*** *In mammals, air is delivered into the lungs via a series of branching bronchi. The ultimate branches terminate in tiny, dead-end, alveolar sacs which consist essentially of a network of fine capillaries separated by a thin layer of epithelial cells from the air. Here oxygen is taken up and carbon dioxide given off. Ventilation of the alveoli depends upon muscular movements of the chest wall and the diaphragm as well as the passive elasticity of lung tissue.*

for each cubic centimeter of air contained in the lung. In man, there are about 300 sq cm of alveolar surface for each cubic centimeter of air. The lungs of a mammal lie in the *pleural cavity*, which is bounded by a "cage" formed by the ribs and the layers of muscle and connective tissue that spread between them. The posterior plane of the pleural cavity is formed by a sheet of muscle and connective tissue called the *diaphragm*. The lungs are not attached to the walls of the pleural cavity, except where the bronchi enter; the lungs cannot pull away from the pleural wall, however, because this would leave a vacuum-filled space, and the pleural wall itself cannot collapse since it is stiffened by the ribs; that is, the lungs are kept plastered against the pleural wall by atmospheric pressure. If the pleural wall is opened, the lung collapses, since the lung tissue is elastic and the pressure is now equal on both sides. If an accident causes penetration of both pleural cavities and both lungs collapse, the respiratory system is crippled and death rapidly ensues.

The ventilation of mammalian lungs is accomplished through the contraction of muscles that raise and lower the ribs relative to the sternum and the vertebral column, thus changing the volume of the pleural cavity. As the volume is increased it must be filled by air rushing in from the outside under the influence of atmospheric pressure. In addition, the contraction of the diaphragm also increases the volume of the pleural cavity and also causes air to enter the lungs. When the chest walls and/or the diaphragm are relaxed, the volume of the chest decreases and air is forced out. The work of inhalation is principally muscular contraction. Exhalation is largely accomplished by the elasticity of the chest wall, diaphragm, and lungs. Why mammalian lungs should lie loosely in a pleural cavity is not entirely clear; perhaps the constant expansion and contraction of the cavity and of the lungs causes less abrasive wear of the delicate lung tissue with this design than would be the case with some other arrangement.

## THE LUNGS OF BIRDS

The high metabolic demands of flight have required of birds the development of many highly efficient mechanisms within their bodies. The respiratory system is almost as highly adapted for flying as are the wings. The lungs are firmly attached to the body wall and arranged so that the movements of the wing muscles help pump air in and out. In addition to the lungs, birds have an elaborate interconnected system of *air sacs*. These sacs connect with the trachea by large tubes that pass directly through the lungs. Other tubes leading from the trachea branch and supply air to many small tubules called *parabronchi*. Each parabronchus is associated with a cylinder of *air capillaries* of which it makes up the core. These air capillaries, about 10 $\mu$ in diameter and about 0.3 to 0.4 mm in length are intimately associated with blood capillaries. Gases are exchanged here. (See Fig. 5-6.) Inspired air may go directly to an air sac or it may pass over the respiratory surface of the bird lung in the air capillaries via the parabronchi. Subsequently, on expiration, the air may also return directly to the exterior or may pass through the parabronchi and thence into the air capillaries and over the gas-exchanging vascular surface. The net effect of this complex system is that during a large fraction of the respiratory cycle, fresh air is moved over vascular surfaces, where respiratory exchange can occur. Thin-walled tubes leading to air sacs ramify into many parts of a bird's body, even into many of the bones. The air sacs are not themselves sufficiently well supplied with small blood vessels to serve as surfaces for respiratory gas exchange with the blood nor are their inner surfaces folded to increase surface area. They are clearly accessory structures that improve the functioning of the avian respiratory system, although their detailed workings have not yet been fully described. Because of their frequent location in long bones, it has often been suggested that the air sacs serve to reduce the weight of the body without reducing the strength of the skeleton. This hypothesis gains strength from the

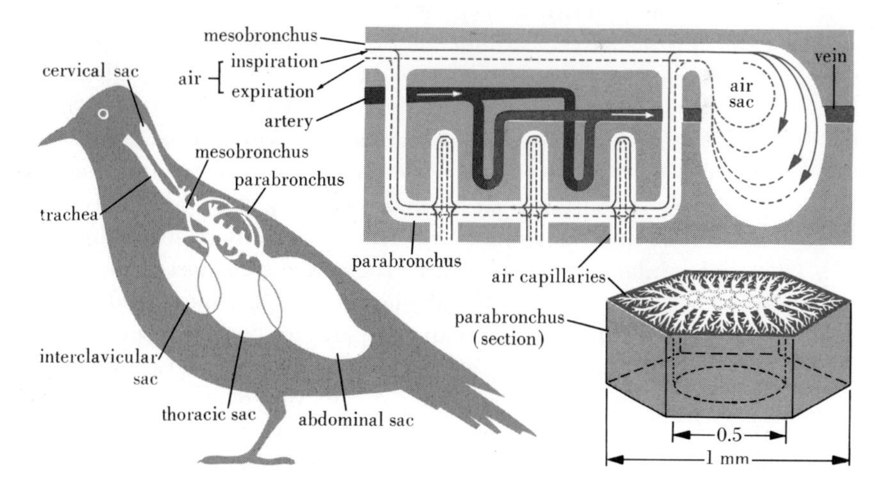


***Fig. 5-6*** *In birds, gaseous exchange in the lungs occurs in tiny air capillaries which surround and take origin from larger air passages called parabronchi. On inspiration, some of the air passes through the parabronchi and diffuses into the air capillaries while some of the air passes through the lungs, unproductively, into the air sacs. On expiration, the air in the air sacs is passed back through the lungs and some of it now enters the parabronchi and the air capillaries. Thus, "fresh" air has access to the air capillaries in both phases of the respiratory cycle. (After Hughes and Zeuthen.)*

fossil evidence that some dinosaurs had air sacs. The air sacs in birds—containing, as they do, atmospheric air, which is at a lower temperature that that of the bird's body—may also be used for temperature regulation. This potential function has been very little studied. It has been observed, however, that the testes in birds are closely approximated to air sacs, which would make possible the local reduction of body temperature. No one knows, however, whether spermatogenesis in birds requires a lower than ordinary body temperature as is known to be the case for spermatogenesis in mammals. (See Chapter 9.) Our ignorance concerning the functioning of the air sacs and the lungs in birds may serve to remind us that many refined physiological mechanisms in animals still await adequate study and analysis. The apparent efficiency of the bird lung also may remind us that mammalian organs are not necessarily the most efficient to be found among living animals.

## *METABOLIC RATES AND FACTORS THAT AFFECT THEM*

Metabolic rates reflect the general activity of the animal in question; sluggish and quiescent animals have lower rates than very active ones. A large number of factors add together to make up metabolism—the sum of all life functions. In man, a quantity known as the *basal metabolic rate* is assessed clinically, usually as a clue to thyroid gland function. The subject's oxygen consumption over a period of time is measured while the subject is in a standard state—fasting, lying quietly in a supine position, in a "comfortable" temperature and humidity, with normal indoor clothing, and at a given time of day (normally on arising). Each of these specifications is important, since any variation may markedly alter the measurement. Digestion is an expensive activity involving muscular and glandular activity as well as the active transport of the products of digestion. The effect on metabolic rate is easily measured and, in fact, may temporarily raise the body temperature. Similarly, posture involves muscular activity. Even sitting quietly may cause a twofold increase in metabolism compared with lying down. Moreover, temperature regulation, by sweating, shivering, or vasoconstriction—which occurs in uncomfortable environments—also causes marked increases in metabolism. Finally, there is a diurnal rhythm of metabolic activity in man, as in almost all living organisms. Because of this rhythm, the time of day of such measurements must be fixed or taken into consideration. The metabolic expense of living that remains to be measured in such a basal state includes respiration; circulation; growth and repair; the production of reproductive cells; glandular secretory activity; the continuing activity of the nervous system, liver, and so forth; and finally the cost of staying alive of every cell in the body. (See Fig. 5-7.) The basal metabolic rate in man is about 41.6 kcal/sq meter of skin surface/hr in a 20-year-old male and about 36.3 kcal/sq meter/hr in a 20-year-old female. On the average, the basal energy requirement for an adult man is 1.3 kcal/min. Such activities as dressing and undressing

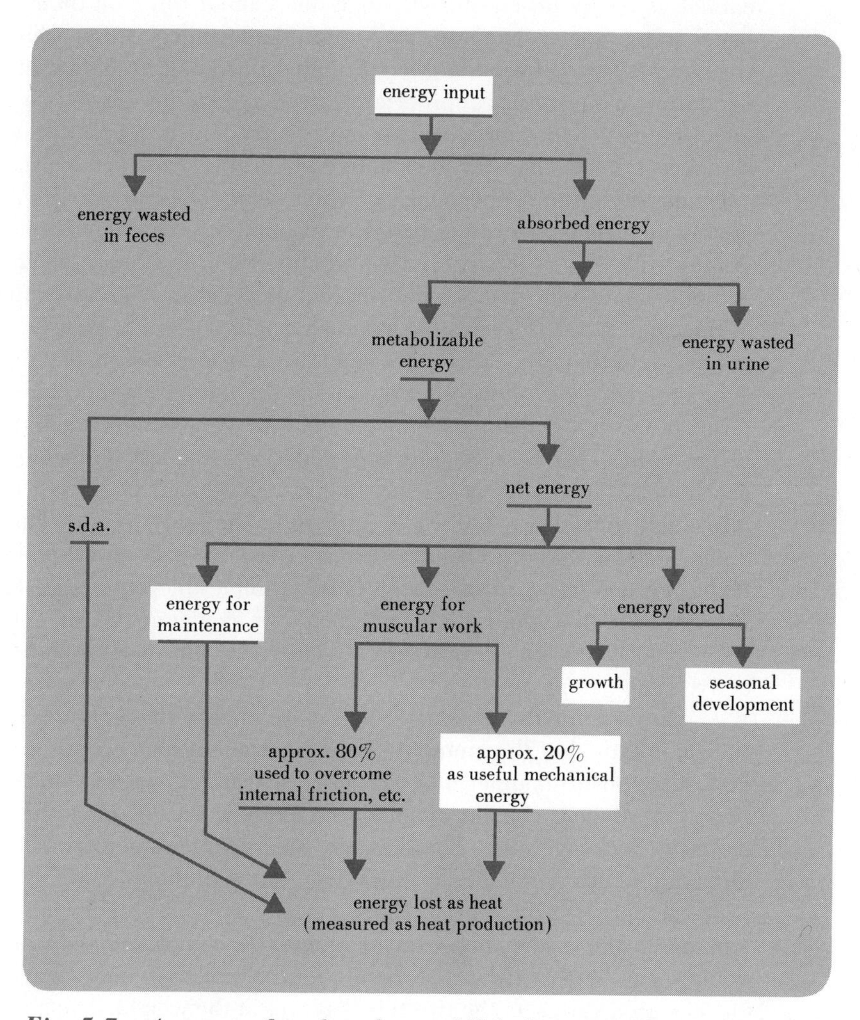


*Fig. 5-7 An energy flow sheet for a goldfish. Input would come chiefly from food but small quantities would also derive from absorbed light, heat, and sound. S.d.a. stands for specific dynamic action or, essentially, the expense of digesting and absorbing food. At death, the energy stored as growth or as fat is transferred to a predator or predators or is slowly released by microbial action. (After P. M. C. Davies,* Comp. Biochem. Physiol., *12:67–79, 1964.)*

require 2.5 to 4 kcal/min. Walking on the level at 4 miles per hour requires about 5.7 kcal/min. Walking up a moderate incline at 2 miles/hr requires about 6 kcal/min. Hard work requires 6.7 kcal/min. or more. Hard exercise such as in a crew race may require up to 20 kcal/min. Hard mental work does not raise the metabolic rate appreciably.

Unfortunately, the basal metabolic rate of animals other than man cannot as easily be measured, since one cannot force on them a truly basal state. Some can, however, be studied while fasting, while relatively quiet, at a chosen and fixed temperature and humidity, and at a fixed time of day. Other appropriately designed studies on poikilotherms give what is called a *standard metabolic rate*. This is generally lower for sluggish animals than for very active ones, the latter probably having substantially altered physiologies, which cannot be turned down completely or turned off even at rest. For example, the mammalian nervous system with its continual activity probably requires greater oxygen consumption than does that of a flatworm, gram for gram. Similarly, the cost of the avian circulatory system even when a bird is not in flight, surely is greater than that of a catfish or a snail. Recent work on the budgerygah, a small parrot, has shown that the metabolic rate in flight (expressed in terms of oxygen consumption instead of heat generation) is about 40 ml $O_2$/gram/hr. That is, a 40-gram bird consumes about 2000 ml of oxygen per hour. At rest, the same bird consumes 320 ml $O_2$/hr. In such a bird, flight muscles make up 26 percent of the body weight. Thus, the flight muscles may be calculated to increase their $O_2$ consumption, in flight, by 2.7 ml/gram of muscle/min. Comparable studies in insects have shown that their flight muscles can consume $O_2$ at a rate of 1.4–7.3 ml $O_2$/gram of muscle/min. These are the most metabolically active tissues known.

Larger animals, of course, use more energy than smaller ones of the same type, but the metabolic rate per gram of living tissue varies in just the opposite manner. For example, the rate of oxygen consumption per gram of body weight is much smaller for an elephant than for a mouse. (See Fig. 5-8.) There is no clear explanation for this fact, although it has often been suggested that metabolic rate is directly proportional to surface area. Since surface area is proportional to the square of the body length (given generally congruent shapes) while volume or weight are proportional to the cube of body length, the surface area varies as the two-thirds power of the weight. The metabolic rate, however, varies among mammals, crustaceans, and fish not as the two-thirds power of weight, but as a somewhat higher power, approximately 0.7–0.75. Some basal functions would not seem to be much more demanding in an elephant than in a mouse. The cost of producing reproductive cells, for example, or of the nervous regulation of regular respiration, or of producing and operating an eye or an inner

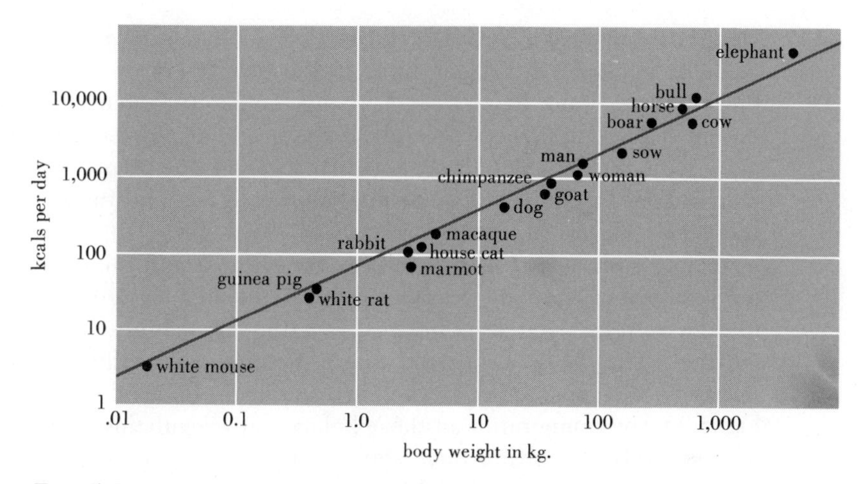


***Fig. 5-8*** *The relationship of body weight (roughly representing volume) to metabolic rate in selected mammals. Note that both axes are logarithmic. The linear relationship on the log-log plot implies that the metabolic rate varies as a power, about 0.75, of the weight. (After Kleiber.)*

ear might reasonably be expected to be only moderately increased with size; certainly far less than proportional. On the other hand, the cost of temperature regulation (depending as it must on the surface area through which heat is lost or gained or on the growth of fur or feathers) would be expected to vary with surface area. The cost of yet other systems (such as the circulatory system, voluntary muscles, skeleton, and perhaps the liver) might be expected to approximate proportionality with weight or to be even greater. Not surprisingly, the observed relationship with size falls in an intermediate zone.

On the other hand, animals below about 1 mm in size usually have a metabolic rate approximately proportional to weight; that is, the intrinsic rate per gram of tissue is nearly constant. One millimeter is also the size above which diffusion is inadequate to supply oxygen to active tissues.

The temperature of an animal's body also has an important effect on metabolic rate. Within the limits that the animal can tolerate, the metabolic rate roughly doubles with every 10° C increase in temperature. Most animals remain nearly at the same temperature as the air or water in which they live. They are called *poikilothermic*, meaning variable in temperature, rather than "cold-blooded," because some of them may have quite a high temperature when in warm air or water. Birds and mammals are exceptions, and are commonly called "warm-blooded." The important distinction, however, is not the warmth but the *regulation* of their body temperature—the ability to hold it close to a constant value despite fluctuations in the surrounding or ambient temperature.

Hence the biological term for birds and mammals is *homeothermic,* meaning constant in temperature. In the hot desert sun, the poikilothermic lizard may have a higher body temperature than the homeothermic camel. In fact, the lizard's temperature, during its active hours, may be as well regulated as that of the camel but by quite different mechanisms. Some reptiles are now known to regulate their temperature quite closely by behavioral means. Thus they may bask in the sunlight (or on hot rocks or sand) to absorb heat and raise their body temperature or they may climb into shady bushes for protection from the sun or to get away from the heat-radiating ground, or they may seek out their burrows below ground where the temperature in hot environments is lower than that of the surface and is relatively stable. (See Fig. 5-9.) The temperature of these behaviorally regulating reptiles may be as closely set during their active hours as that of some mammals or birds. Contemporary reptiles probably represent some of the steps

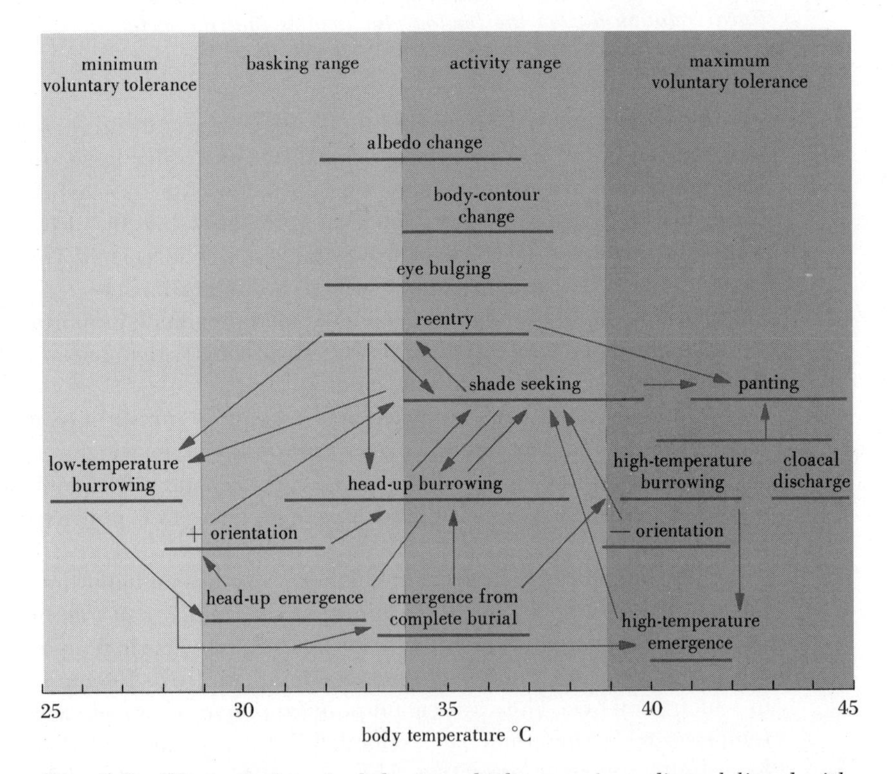


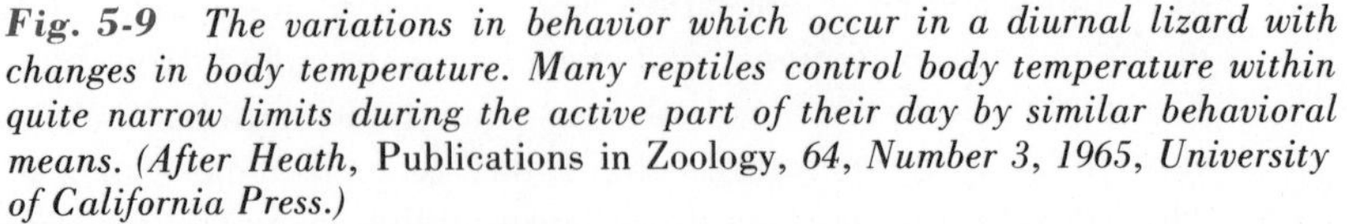
***Fig. 5-9*** *The variations in behavior which occur in a diurnal lizard with changes in body temperature. Many reptiles control body temperature within quite narrow limits during the active part of their day by similar behavioral means. (After Heath,* Publications in Zoology, *64, Number 3, 1965, University of California Press.)*

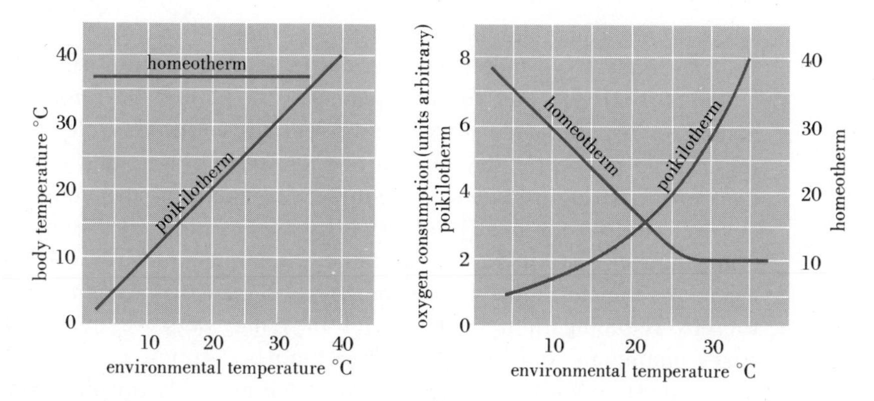


***Fig. 5-10*** *The changes in body temperature and oxygen consumption (an index of metabolic rate) in a typical homeotherm and poikilotherm with changes in ambient temperature. (After Bartholomew.)*

that at least two lines of other reptiles, the ancestors of mammals and birds, passed through before homeothermy was achieved. At other times —night or winter, for example, depending on the species—a reptile's temperature may approximate that of its environment. Probably other terrestrial groups such as insects will eventually be shown to have some parallel degree of regulation. Moths have already been shown to regulate their body temperature. Regulation of colony temperature by bees is another interesting example. (See *Behavior* in this series.)

Regulation, by behavior or otherwise, is far more difficult for aquatic animals since heat is so much more rapidly transferred by water than by air. Elements of temperature regulation, for part of the body mass, have recently been shown in large oceanic fish, including sharks and tuna. Our knowledge of what degree of behavioral and physiological regulation has been achieved by aquatic reptiles is still fragmentary. In the case of aquatic mammals and birds, there are many highly specialized adaptations such as thick layers of blubber under the skin and countercurrent heat-conserving mechanisms.

The metabolic rate of poikilothermic animals falls rather rapidly with decreasing temperature, approximately to half its previous value for a temperature drop of 10°C. This, of course, renders many such animals sluggish or inactive and may restrict their active periods to certain seasons of the year or to certain times of the day. Homeothermic animals react metabolically in just the opposite manner. When exposed to cold air or water, they generate more heat to maintain the normal temperature of their bodies. (See Fig. 5-10.) Their oxygen consumption increases at lower ambient temperatures as metabolic energy is spent for heat production. This added expense for heat production is held down as much as possible by certain other structural and physiological adaptations. Those animals, for example, that are routinely exposed to cold normally have increased thermal insulation in the form of fur or feathers. Both fur and feathers can also be fluffed up for increased insulation or, if cold occurs during regular seasonal cycles, heavier fur may be grown for that period of time. Both fluffing and growing of heavier coats require energy, of course. Animals may also conserve the energy required for heat production by preventing heat loss by means other than insulation. Not uncommonly, mammals or birds will assume a posture that minimizes the exposure of their extremities to the cold, thereby reducing their heat loss, or they may seek shelter. In a few cases, mammals and birds can allow their body temperature to drop a fair amount without any metabolic reaction and only invoke extra heat production when the temperature drops past a particular, crucial point. (See Fig. 5-11.)

Homeotherms must also expend energy to stay cool when the ambient temperature rises above that of the body, and equally ingenious mechanisms have evolved for this. Basically, cooling mechanisms usually depend on the evaporation of water, which is a heat-absorbing transformation. Many mammals sweat; some pant (evaporating water from the surface of the respiratory tract); and still others (some small rodents) drool and spread the saliva onto their skin and fur from which it then evaporates. Cooling can also be facilitated by increasing surface area (provided the body is warmer than the surrounding air) as is apparently the function of elephants' ears, by increasing the blood flow to the skin (thereby carrying heat from the interior to the surface) or by moving to a cooler microenvironment, such as a burrow or cave, or into water. Surface area (exposed to the environment) can be increased by changes in posture.

Several kinds of mammals and a few species of birds exhibit a special metabolic state in which they sometimes allow their body temperature to fall to that of the surrounding air for varying periods of time. This is usually called *hibernation,* although it may occur at seasons other than winter, and it serves to reduce, considerably, the animal's metabolic rate and hence its consumption of food. Hibernation, or a functionally similar state, usually occurs when food is scarce. The animal then usually utilizes reserves of fats stored within its body. At a reduced body temperature, the lowered metabolic rate conserves these stores and permits survival over a longer period of hard times. Naturally there must be many associated physiological adaptations for hibernation such as providing for reawakening at the proper time and providing for some degree of protection from predation or from weather. There are many varied designs of hibernation. In some cases, there may be a rather continual state of lowered metabolism for months. In other cases —hamsters, for example—the animals may awaken briefly to eat

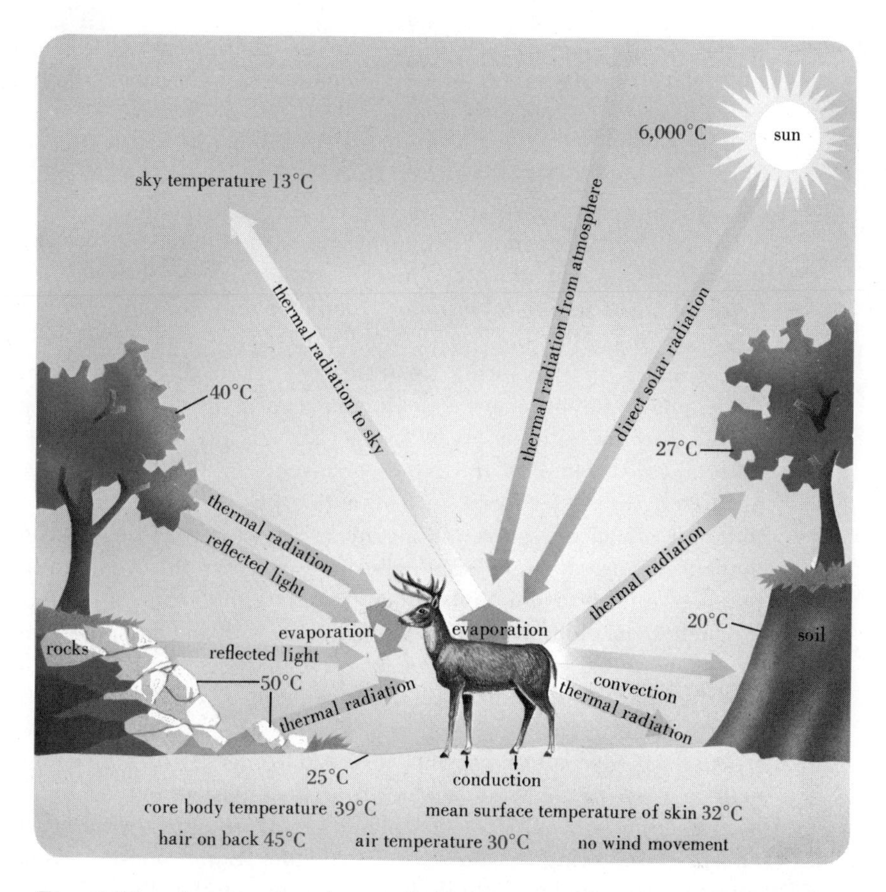


**Fig. 5-11** *An animal exchanges heat with many aspects of its environment. The direction of exchange by radiation of course depends upon body (and local, surface) temperature as well as the temperature of the ground and nearby vegetation, the height of the sun, cloud cover, and the radiating qualities of the skin (or antlers) and of the terrain. The animal may be absorbing heat from the sun while radiating to the shaded ground beneath it. Heat may also be gained or lost by convection which is strongly influenced by wind or by conduction which depends upon the medium (air, water, or an insulating sleeping bag, for example) and area of contact. Heat is also lost by insensible evaporation from the body surfaces (especially from the skin and respiratory tract) and by sweating in many mammals. Less important increments of heat are gained or lost with food, water, urine, and feces. (Adapted from G. Bartholomew in* Animal Function: Principles and Adaptations, *Macmillan, 1968.)*

stored food or to assess the environment. Perhaps one of the most striking examples of adaptation to hibernation has been revealed by extensive studies of temperate-zone bats. In several species of such bats, copulation occurs in the fall before hibernation but there is, in some, delayed fertilization, with storage of the sperm until spring in the female genital tract, while in others there is prompt fertilization but delayed implantation and development. In both cases, the function appears to be to spare the female the metabolic burden of nourishing a fetus during the difficult winter season, while taking advantage of the vigorous good health of the male in the fall for sperm production.

In many insects, a state of suspended metabolism called *diapause* occurs. Like hibernation, diapause spares metabolic energy during a difficult season and allows the insect to re-emerge when times are again favorable. But unlike hibernation, diapause may continue even at high temperatures. Diapause also synchronizes the emergence of males and females in the spring regardless of when they completed their development the previous year. Diapause normally characterizes a stage in the life cycle of a given insect and is, presumably, genetically determined. Examples are known of insects in which diapause occurs as an egg (just fertilized or containing an embryo), a larva (many different stages), a pupa, or an adult. Examples are even known of diapause characterizing every alternate or every third generation (depending on life span and climate, apparently). The spruce sawfly, in Quebec, has but one generation per year, with an obligatory diapause. In Connecticut, however, three generations occur per year and these may or may not diapause. In the mosquito, *Aedes dorsalis*, there are several generations per year, but only those eggs laid in the fall diapause. Diapause is broken, and the insect returns to its active metabolic state, in response to physical clues from the environment—cold followed by warmth, light-dark cycle (photoperiod) changes with season, or moisture following dryness, for example. Some moths even have translucent "windows" built into the skins of the diapausing pupae. These windows apparently allow the pupal brain to respond to changes in day length. The metabolic rate of a diapausing silkmoth pupa may be of the order of 10–20 $\mu$l $O_2$/gram wet wt/hr while that of the silkmoth pupa emerging from diapause may be 100–200 $\mu$l $O_2$/gram wet wt/hr. The considerable saving involved is apparent.

***ANAEROBIC METABOLISM*** Important exceptions are found to the general rule that oxygen is required for all animal metabolism, for there are animals that live in environments where oxygen is virtually absent. Often mud contains sufficiently dense populations of bacteria feeding on organic debris to use up all the oxygen; another environment almost devoid of oxygen is the intestinal tract of mammals. Some of the nematodes living in mammalian intestinal tracts obtain energy from fats and carbohydrates by breaking them down only part way, which can be done without oxygen. They give off as waste products substances such a lactic acid ($CH_3CHOHCOOH$), which in most animals would be oxidized further to carbon dioxide and water.

The partial breakdown of carbohydrates by the anaerobic nematodes is quite similar to the first steps in the breakdown of glucose in animals that do require oxygen, and is called *glycolysis*. In many cells, glycolysis can proceed at a faster rate than the subsequent steps, which lead to the complete utilization of the energy available in glucose. For example, muscle cells often work at a rate greater than can be accommodated by their ability to take up oxygen, which is limited by the oxygen delivery capacity of the circulatory system. Under these conditions, lactic acid accumulates and the muscle is said to have acquired an *oxygen debt*. When the period of strenuous work ceases, the muscle continues to use more oxygen than it would normally require at rest. This process is called "paying off" the oxygen debt; athletes will recognize this as the period of heavy breathing that follows immediately after a period of maximum exertion. During this time, the accumulated lactic acid passes through the remaining oxidative steps in the sequence of cellular metabolic reactions that had previously been blocked by the inadequate supply of oxygen.

Many microorganisms obtain much or all of their metabolic energy, without using oxygen, by glycolysis or similar methods of partially breaking down food molecules. In fact, there is good reason to believe that the earliest living organisms were limited to this type of metabolism by the virtual absence of free oxygen in the earth's primeval atmosphere. We believe that only later, when green plants had produced vast quantities of oxygen as a by-product of photosynthesis, did fully *aerobic* metabolism become commonplace in animal cells. The much greater amount of energy obtainable from each molecule of food has long since made complete oxidation to carbon dioxide and water the most advantageous system and, hence, the most widespread type of animal metabolism.

## FURTHER READING

Brett, J. R., "The Swimming Energetics of Salmon," *Scientific American*, 213 (2): 80–85, 1965.

Comroe, J. H., Jr., "The Lung," *Scientific American*, 214 (2): 56–68, 1966.

Davson, H., *A Textbook of General Physiology*, 3d ed. Boston: Little, Brown, 1964.

Dawkins, M. J. R., and D. Hull, "The Production of Heat by Fat," *Scientific American*, 213 (2): 62–67, 1965.

Florey, E., *An Introduction to General and Comparative Animal Physiology*. Philadelphia: Saunders, 1966.

Gordon, M., *Animal Function: Principles and Adaptations*. New York: Macmillan, 1968.

Hoar, W., *General and Comparative Physiology*. Englewood Cliffs, N.J.: Prentice-Hall, 1966.

Hughes, G. M., *Comparative Physiology of Vertebrate Respiration*. Cambridge, Mass.: Harvard University Press, 1963.

Irving, L., "Adaptations to Cold," *Scientific American*, 214 (1): 94–102, 1966.

Johansen, K., "Air-breathing Fishes," *Scientific American*, 220 (1): 102–111, 1968.

Kayser, C., *The Physiology of Natural Hibernation*. New York: Pergamon, 1961.

Lees, A. D., *The Physiology of Diapause in Insects*. New York: Cambridge, 1955.

Moore, J. A. (ed.), *Physiology of the Amphibia*. New York: Academic Press, 1964.

Morosovsky, N., "The Adjustable Brain of Hibernators," *Scientific American*, 218 (3): 110–118, 1968.

Romer, A. S., *The Vertebrate Body*, 3d ed. Philadelphia: Saunders, 1962.

Ruud, J. T., "The Ice Fish," *Scientific American*, 213 (5): 108–114, 1965.

Slonim, N., and J. L. Chapin, *Respiratory Physiology*. St. Louis: Mosby, 1967.

Taylor, C. R., "The Eland and the Oryx," *Scientific American*, 220 (1): 88–95, 1969.

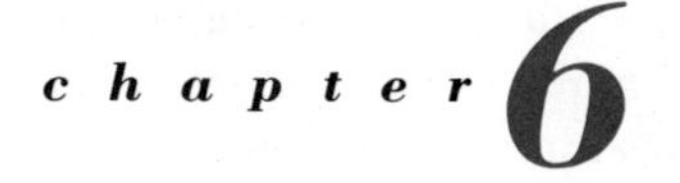

# Motility, Contractile Tissues, and Effector Organs

Living protoplasm is seldom static. In addition to the agitation of cytoplasmic particles by random molecular bombardment, known as Brownian movement, there are grosser cytoplasmic flows and distortions of cell shape that depend on energy derived from cellular metabolism. In ordinary cells, these movements may be rather slow. Time-lapse motion pictures, in which a photograph is taken every several seconds or minutes and then projected at 16 frames per second, demonstrate large, but irregular, movements and shape changes of even such major cell components as the nucleus and the mitochondria. In some cells, the movements may be quite rapid. Sometimes the semiliquid cytoplasm flows, usually in a rough circuit, in such a way that a given particle may be carried back past its starting point again and again. In amoebae, or in the amoebocytes of multicellular animals, *cytoplasmic streaming* can be seen while lobes or strands of protoplasm are pushed out and retracted in locomotion and in the engulfment of food particles. In muscle cells, protoplasmic movements are concentrated and organized so that well-defined contractions can occur at the proper times and places within an animal's body.

Protoplasmic movements appear to depend on the properties of fibrous aggregates of special protein molecules. Metabolic energy is used for these movements, and much the same biochemical processes, and probably very similar protein molecules, are involved in the mild, slow, and occasional contortions of connective tissue cells as in the rapid and forceful shortening of highly specialized muscle cells. In all cases, the energy to power the movement comes originally from the oxidation of food molecules. *Adenosine triphosphate* (ATP) appears to be the intermediate vehicle for energizing the interactions between proteins which cause movement.

The movements of muscle cells have been studied in much more detail than any of the other types of movement. We know that muscle cells—and probably some other cells that produce slower, amoeboid movements—contain a protein complex called *actomyosin.* Actomyosin contains long filamentous chains of a globular protein called *actin,* combined with aggregates of a much larger protein called *myosin.* Myosin is known to be an enzyme which can release the energy carried by ATP. Although these are often referred to as "contractile proteins," recent evidence indicates that the interaction between actin, myosin, and ATP, which causes the contraction of muscle cells, does not actually involve the shortening of actin or myosin molecules. The detailed mechanism for this interaction is a challenging puzzle for modern biochemists.

In some other cells, movement is associated with the presence of relatively rigid structures called *microtubules* instead of the thin filamentous chains of actin found in muscle. The biochemistry of movement in these cells has only recently begun to be studied. The movements of chromosomes during cell division and the movements of the bodies of some protozoa appear to involve microtubules, but the best-known examples are the movements of *cilia* and *flagella.* Cilia and flagella are hairlike organelles that project not only from the surfaces of many microorganisms, but also from the cells forming internal or external surfaces of many multicellular animals. Where they are external, the beating cilia may serve to move the entire animal as in some flatworms and in the larvae of many invertebrate animals. A beating flagellum is commonly found on the motile gametes of plants as well as animals. More common, however, are ciliated "internal" surfaces over which water currents are set up by the coordinated beating of thousand of individual cilia. Such ciliary systems vary, from those that generate currents flowing over the gills of clams to bring water to the respiratory surfaces and food to the mouth to those that line our own trachea or windpipe to move a current of mucus upward toward the pharynx, thus carrying away from the lungs particles of dust and foreign matter.

The integrated, rhythmic movements of cilia and flagella appear to require control mechanisms within the organelle involving a significant amount of interaction between different regions of the organelle. Thus it is not surprising that cilia are often involved in the detection of sensory stimuli. The cilia of the *hair cells* of the inner ear of vertebrates do not move spontaneously, but are moved relative to the cell by gravity, acceleration, or sound waves, and it is this relative movement which is detected and interpreted by these sense organs. Similar ciliary movement detectors are found in many invertebrates. In addition, the light-detecting cells in the eyes of many kinds of animals, including vertebrates, contain modified cilia, which no longer move. Much needs to be learned before we understand either the movement or the sensory-detecting capabilities of cilia and flagella. The structure and function of cilia are further described in *Microbial Life* in this series.

## THE STRUCTURE OF MUSCLE CELLS

In multicellular animals, movement is one of the major body functions most likely to exemplify the division of labor among specialized cells. In coelenterates, as well as in more highly organized animals, some cells having a high concentration of contractile protein fibrils are called *muscle cells.* Usually they are elongate and their ends are attached in some fashion to the surrounding tissues so that their shortening pulls on part of the animal relative to another part. In such highly organized animals as arthropods, mollusks, and vertebrates, the muscle cells may become packed with the contractile protein complex called *actomyosin.* The structure of muscle cells varies widely from one type of animal to another and in the different organs of a single, complex animal. The simplest are long, spindle-shaped cells with the contractile protein molecules oriented mainly parallel to the axis of the cell but not readily distinguishable from the rest of the cytoplasm by ordinary microscopic examination. The *smooth muscle cells* in the wall of the gut and in the walls of the blood vessels, in vertebrates, are of this simple type.

In many highly organized animals, muscle cells form a *syncytium,* where a large volume of protoplasm containing many nuclei is contained within a single cell membrane. Such a syncytial unit is called a muscle fiber; it usually achieves more rapid or vigorous contractions than does a smooth muscle cell. In some cases, but by no means all, the actomyosin is organized on the macromolecular level so that elements of the molecular structure are regularly spaced throughout the syncytium. Such an arrangement produces under the light microscope a series of visible, regular cross striations because of which these muscles are called *striated.* Most of the "muscles" of vertebrate animals are this type, including all of those, called skeletal muscle, that are attached to parts of the skeleton and serve to move one part of the skeleton relative to another. Many invertebrates, including even some coelenterates, also contain some striated muscle. Among invertebrates several other types of muscle also occur which can be distinguished, in some cases, by their microscopic appearance and, in some cases, by their physiology. The hearts of vertebrate animals contain a special type of striated muscle, *cardiac muscle*, which is characterized by a series of branching and interconnected muscle fibers which have thin transverse partitions representing the walls of individual cells. (See Fig. 6-1.)

The organization of the actomyosin in striated muscle has recently been revealed by the combined use of the electron microscope and selective biochemical analysis of the different elements in the striated muscle fiber. The typical ultrastructure of a striated muscle is shown diagrammatically in Fig. 6-2; the identity of the molecular components has been established by first extracting samples of muscle with reagents known to remove selectively the actin or the myosin and then making electron microscope preparations that show the absence of the material

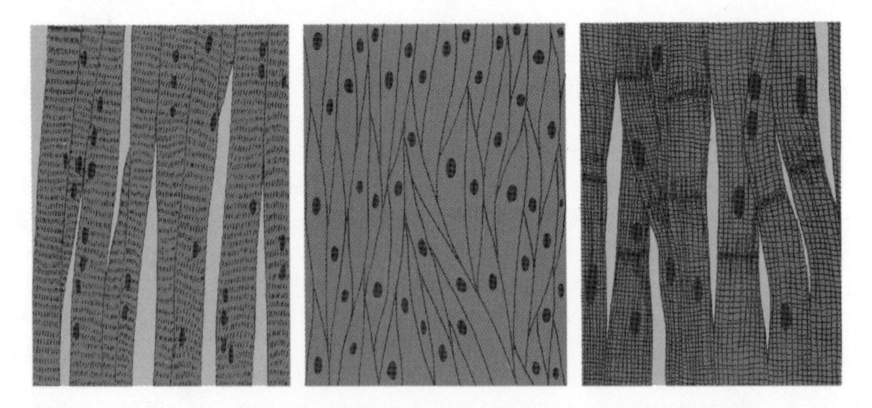

**Fig. 6-1** *Common types of muscle cells found in vertebrates. Smooth muscle* (center) *consists of separate cells. Striated muscle* (left) *is syncytial, containing many nuclei. The same is true for the electroplates of electric fish which are derived from striated muscle. See Fig. 6-9. Cardiac muscle* (right) *shares some features with both smooth and striated muscle.*

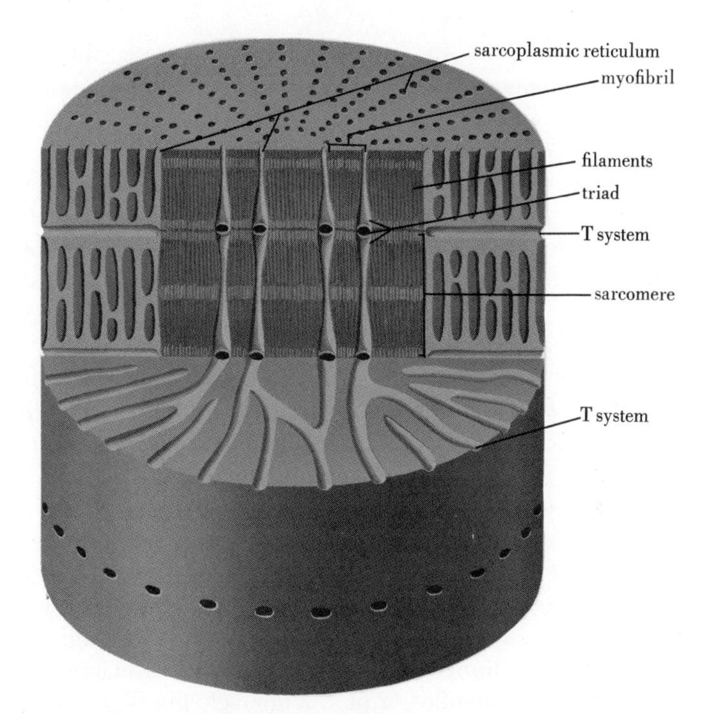


*Fig. 6-2 The electron microscope has revealed a great deal of fine structure in vertebrate striated muscle. The T system is continuous with the surface cell membrane and is believed to function in conducting the muscle action potential into the interior of the syncytium. The enormously redundantly folded intracellular sarcoplasmic reticulum (associated with energy conversion) intersects the T system in regions of recognizable pattern called "triads". The repetitive subunits of muscle fibers—myofibrils, filaments, and sarcomeres—appear to have functional significance, not all aspects of which are yet clearly identified. (After Porter and Franzini-Armstrong.)*

so removed. Similar techniques continue to be used to identify and explore the dynamics of the energy transformations involved in muscle activity.

The surface membrane of a muscle cell is an important part of its structure, for it is here that electrical events, similar to those that occur at the surface membrane of a nerve cell (see Chapter 10), occur to control the contraction of the cell. A resting muscle cell remains relaxed even though it contains ATP and actomyosin. Apparently the release of free calcium ions into the cytoplasm is required before the ATP and actomyosin can interact to cause contraction. In a small muscle cell, the electrical events that occur at the surface membrane when the cell is stimulated can apparently cause a direct release of calcium ions into the cytoplasm, causing the cell to contract. In large, striated muscle cells, the membrane is too far from the interior of the cell for this to happen. These cells have a system of intracellular membranes, consisting in part of infoldings of the cell surface, known as the *sarcoplasmic reticulum.* The electrical events at the surface of these cells are apparently transmitted, in a manner that is not yet well understood, to the membranes of the sarcoplasmic reticulum, which in turn control the release and recovery of calcium ions. Rapidly contracting muscle fibers have a particularly extensive sarcoplasmic reticulum.

## FUNCTIONAL SPECIALIZATIONS OF MUSCLE CELLS

Although single muscle cells only shorten, aggregates of such cells in muscles are called upon to perform a variety of mechanical movements and to produce various degrees of force over shorter or longer periods of time. They are also elastic structures, quite apart from their contractile function; they have some of the properties of a rubber band. Most muscles have a given resting length. If stretched they resist by an elastic force that increases as they are stretched. Part of this elastic quality resides in the cell membranes or in the thin sheets of elastic connective tissue that surround bundles of parallel muscle fibers. But many smooth muscles exhibit different resting lengths from time to time. Such differences in elasticity and resting length seem to be highly correlated with the activity the muscle is designed to undertake. Smooth muscle cells of the wall of vertebrate gut, for example, have to work effectively against different volumes of food or against markedly different effective calibers of the lumen. They must yield to large volumes but must also be competent to contract against such volumes effectively in order to churn the food or to move it. In order to do so, a gradation of resting lengths is desirable. Skeletal muscle, on the other hand, is more closely designed for functioning in a fixed geometry (fixed by jointed, skeletal elements). For this role, a fixed resting length seems effective.

One specialized type of muscle is found in instances in which the animal must maintain a steady tension for long periods of time. The muscle that may hold an oyster's shell closed for hours when the animal is subjected to danger or to adverse conditions is a good example of this type. It is competent to resist powerful attempts to open the shell from outside. There appears to be a "catch" element that allows these muscles, once they have contracted, to resist stretching with little or no further expenditure of energy. Predatory starfish, however, may fix their suction-cup-like tubular feet to such an oyster shell and pull for minutes or hours in a persistent fashion until they have separated the two valves enough to allow passage of their everted digestive system by which they digest the mollusk in place.

In many animals, including the vertebrates, there are tubular organs such as the intestinal tract and blood vessels which often function best if their walls are always in a state of constrictive tension. The degree of constriction must vary continually as activities and needs change. Elastic connective tissue will not suffice since changes in caliber of the lumen are required. In these situations, smooth muscle cells are arranged in layers, often alternately longitudinal and circular; their slow but steady contractions serve to maintain the *tone* or state of tension of the visceral wall and caliber of the lumen. In vertebrates, smooth muscle is found, among other places, in the walls of the digestive tract, including the pancreatic and hepatic ducts; in the walls of the respiratory passages, such as the trachea and bronchi; in the walls of blood vessels larger than capillaries; and in the walls of the ureters and the urinary bladder. Here smooth muscles may churn and/or move material to be digested in a regular fashion (peristalsis), regulate the freedom of movement of air in and out of the lungs, maintain blood pressure, shunt more or less blood to given organs or organ systems (as in the case of conserving body heat by constricting the blood vessels of the skin), move urine from the kidneys to the bladder, and so on. In mammals, tiny smooth muscles are arranged just below the skin to pull on the base of each hair, raising the hairs and increasing the thickness of the fur for increased thermal insulation or for protection of the central nervous system while they are engaged in aggressive behavior or when frightened (the hair is often raised along the spine) or to change the silhouette (apparently a device for communicating emotion or readiness to attack in many species). Muscles that maintain a steady tension are called tonic muscles, and the state of sustained constrictive tension is referred to as *muscle tone.*

At quite another extreme of muscle function are the striated skeletal muscle fibers. Among their main functions are to move the animal quickly or to produce rapid movements of its limbs and to permit fine gradations of movement. Striated muscles can and do contract fully within a fraction of a second after being excited by the motor nerves that supply them. They can also relax quickly so as to permit rapid oscillatory movements when these are needed. Their alternate contraction and relaxation in a coordinated fashion leads to such complex movements as swimming, running, or flying. The wings of small birds are moved back and forth rapidly—up to 50 times per second in hummingbirds. The rapid oscillations of insect wings (up to several hundred times per second) are often caused by a special kind of muscle, called *fibrillar muscle.* These muscles, when attached to a load having the elastic and inertial properties normally provided by the insect's wings and skeleton, can contract and relax at a much faster rate than that of the electrical events that occur at the surface membrane.

Individual striated muscle fibers usually contract in a unitary fashion, following what is often called the "*all-or-none*" *principle.* These fibers either exhibit the full contraction of which they are capable under the prevailing conditions or else do not contract at all. This all-or-none reaction stems from arrangements within the syncytial fiber that assure its total activation each time the nerve cell adequately stimulates the fiber. The muscle as a whole—containing many individual muscle fibers—may have a very large number of graded degrees of contraction. Each fiber is governed by the all-or-none principle, but all need not contract together. The muscle fibers are usually arranged in groups or *bundles*, of different sizes in different muscles; each such group may contract individually. Thus these voluntary muscles can usually produce anything from a very small movement or tension to a very large one—as is the case, for example, of the movements and the strength of the movements our fingers can make.

Cardiac muscle must contract and relax rhythmically in order to move the blood through the heart systematically and to pump the blood out to the body. Since vertebrate animals are dependent on a continuous oxygen supply to their cells, heart muscle must beat without any appreciable pause for the animal's entire life span. The rate, of course, varies. Cardiac muscle, and some types of smooth muscle, are intrinsically rhythmic; even an isolated piece of heart muscle will contract and relax spontaneously at intervals. The contraction of different parts of the heart must be temporally coordinated so as to move the blood smoothly from chamber to chamber. The structure of vertebrate heart muscle is one of branching and interconnecting fibers, very much like striated muscle in other respects. In a functional sense, the normal heart behaves as a unit and the contraction of a heart or one of its single chambers is ordinarily all-or-none. That is, when a heart contracts, it does so fully, producing as much contractile force as it can under the given conditions. These variable conditions include blood volume, body temperature, and $O_2$ tension. Although cardiac muscle consists of a series of individual cells, separated by cell membranes visible under the electron microscope, the spreading wave of contraction is not stopped or even appreciably slowed by these membranes. Normally, the beat initiates in specialized tissue located at the point where venous blood returning from the body enters the heart. In fish, this is a portion of an initial, large, thin-walled chamber, the sinus venosus. (See Fig. 7-3.) In mammals, there remains, as a vestige of the sinus venosus, a *node* of specialized muscle tissue called the *sine-atrial* or *SA node*, which is apparently more irritable than the rest of the cardiac muscle or has a lower threshold for spontaneous activity. (See Fig. 6-3.) Such a node is called a *pacemaker.* In any event, normal contractions originate here and spread over the two atria. (See Chapter 7.) At the

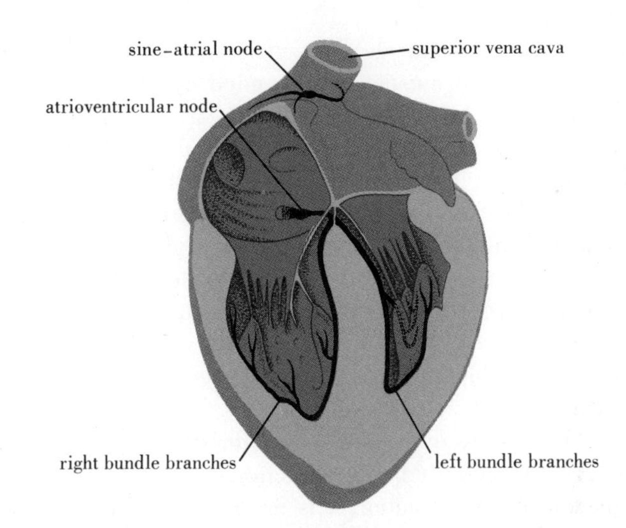


***Fig. 6-3*** *Sequential and synchronized contraction of the mammalian heart depends upon two nodes of specialized muscle tissue, the sine-atrial (SA) node and the atrioventricular (AV) node, and upon other specialized muscle tissue, the bundle branches or bundles of His. The contraction begins at the SA node and is carried over the surface of the atria muscle, causing atrial contraction and firing of the AV node. The AV node then triggers waves of excitation in the bundle branches which cause ventricular contraction. These various waves, as recorded from conventional sites on the body surface by a potentiometer, make up what we know as an electrocardiogram.*

atrio-ventricular junction, a second node, the *atrio-ventricular* or *AV node*, on being triggered by atrial contraction, sets off ventricular contraction. A specialized bundle of cardiac muscle tissue, the *bundle of His*, rapidly conveys the signal to contract from the atrio-ventricular node to all parts of the ventricles. When these nodal trigger points or the bundle of His are damaged, arrhythmias or asynchronous contraction results. The rate at which the sine-atrial node beats is influenced by both the parasympathetic and sympathetic nervous systems and by circulating adrenalin. (See Chapter 10.) In recent years, electronic pacemakers have been developed for implantation in the hearts of patients whose own cardiac conduction system is faulty.

## THE ORGANIZATION OF MUSCLES

Whole animals do not move about by the activities of disorganized clumps of muscle cells. Muscle activity is coordinated by the nervous system, so groups of muscles act together. Each individual muscle is itself a harmonious community of contractile units—whether these be smooth muscle cells, striated muscle fibers, or the branching fibers of cardiac muscle—surrounded by thin but tough sheets of connective tissue. In one of the simplest arrangements, the axes of all the cells or fibers are parallel so that they all pull in the same direction. In many muscles, however, the need is not for a simple pull in a single direction, but for some parts of a large muscle to pull in slightly different directions from other parts or to develop a force greater in one part than in another. Many muscles are called upon to develop tension but not to contract over any great distance. Often their fibers are arranged at a considerable angle to the direction of pull, so that as they shorten and thicken, the whole muscle shortens only slightly but with great force.

One of the important characteristics of muscle cells is that they can exert force only in one direction—when they are contracting. Once a muscle has contracted, it is not able to extend itself. The cycles of contraction and extension that are required for locomotion and other effective movements almost always require two or more muscles. The action of these muscles is coordinated by some form of skeleton, so that the contraction of one muscle can be reversed by contraction of other muscles. Muscles which interact in this way are often described as *antagonistic.*

The clearest examples of antagonistic muscles are found in the limbs of vertebrates or the various jointed appendages of arthropods. (See Fig. 6-4.) Almost every joint of a vertebrate or of an arthropod is bent or flexed by one or more *flexor* muscles and is straightened or extended by one or more *extensor* muscles. The combination of jointed

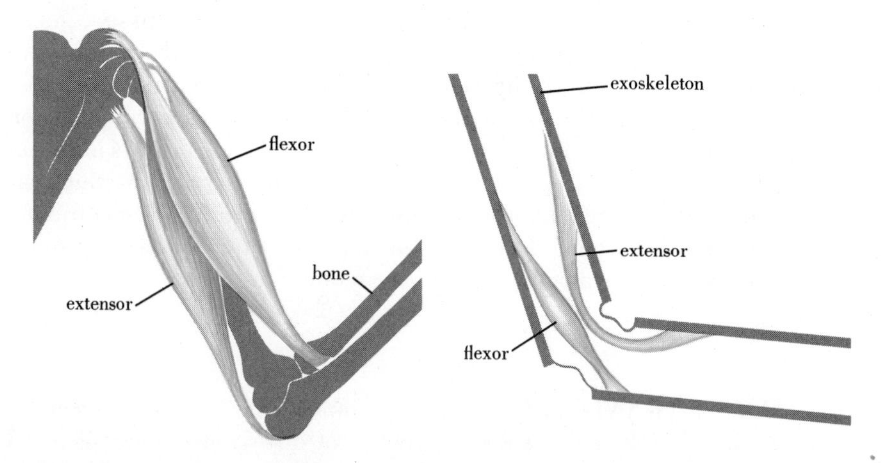


**Fig. 6-4** *Typical and simple arrangements of opposed muscles controlling joint position and limb movement or posture in a mammal* (left) *and in an arthropod* (right). *Much more complex arrangements also occur.*

skeletons and discrete muscles found in vertebrates and arthropods forms a basis for accurate and reproducible movements, which allow these animals to develop sophisticated patterns of behavior. Normally, both the extensor and flexor muscles will act together, but in different degrees; thus, fine adjustments of movement and tension are possible. In addition, antagonistic muscles are stimulated by the nervous system in an automatically coordinated fashion, so when one contracts the other one is inhibited and partially relaxes. Muscle antagonism is not a disorganized competition, but a coordinated, reciprocal arrangement whereby both members of the pair maintain a reasonable tension on a joint; while one relaxes, the other shortens to move the joint.

In a few cases, the contraction of a muscle may be antagonized not by other muscles but by the elasticity of elastic connective tissues. The lens musculature of the mammalian eye is a complicated example of this sort. The ideally spherical lens is normally held in a somewhat flattened condition by the tension transmitted to it by a band of connective tissue fibers attached to the lens capsule. With the lens in this shape, the eye is focused for distant vision. Contraction of the *ciliary muscle* (a ring of smooth muscle cells) releases the tension on the distal end of the elastic connective tissue fibers, thus allowing the lens to round up for focusing on near objects. With the loss of elasticity in the lens itself with age, the lens no longer assumes a spherical shape when released from tension and farsightedness results. Elderly people, for example, must usually hold a book at arm's length to focus on the print unless they use eyeglasses. Other examples of elastic antagonism are found in the shell hinges of bivalve mollusks and the body wall of nematodes.

In many animals, the body wall and the intestinal tract and many other hollow tubular organs contain two layers of muscle; in vertebrate animals, except for the body wall, these are usually smooth muscle cells. One set is arranged in a circular fashion so that its contraction constricts the vessel or squeezes the contents; the other set is arranged longitudinally at right angles to the circular muscles—that is, parallel to the axis of the tube. When these longitudinal muscles contract, the tubular organ shortens and thickens. This arrangement is also found in the cylindrical body wall of coelenterates, the body walls of annelid worms, and in the intestinal tracts of all highly organized animals. The alternation and coordination of contraction of such circular and longitudinal muscles changes the shape of these tubular cavities in a variety of ways. A wave of contraction of the circular muscles, for example, may proceed progressively along the tube, and this process, known as *peristalsis*, forces the contents in one direction. A special type of circular muscle, forming a *sphincter*, separates two segments of a tubular organ or controls its entrance or exit. Examples of such would be the anal sphincter at the posterior end of the intestinal tract or the pyloric sphincter, which regulates the flow of gastric contents into the small intestine. When the circular and longitudinal muscles of a tubular organ or a soft-bodied animal surround a closed cavity, the muscles are all mutually antagonistic. Since the volume of the cavity cannot decrease, the contraction of one muscle must cause the elongation of other muscles. Such an arrangmenet of antagonistic muscles is referred to as a *hydrostatic skeleton.* The earthworm is a good example of an animal which has this type of hydrostatic skeleton; contraction of its body wall muscles can cause locomotion, even though the familiar type of rigid skeleton is not present. The hydrostatic skeleton is not uncommon even in groups of animals that also have rigid skeletons. The tube feet of echinoderms, for instance, operate on this principle.

There are other muscles having complex arrangements of individual fibers or groups of fibers. The human tongue is a familiar example, as is the foot of mollusks, such as clams and snails. These organs, consisting of muscle and connective tissue, can assume a great variety of shapes by selective contraction of separately innervated fibers oriented in several different directions.

## THE ACTIVATION OF MUSCLE CELLS

The contraction of a muscle is almost always controlled by the animal's nervous system. The nature and functioning of nervous systems will be considered in Chapter 10. *Motor nerve cells,* or *motor neurons,* carry signals that stimulate muscle cells to contract or inhibit them from contracting. In the most familiar case, the typical voluntary skeletal muscles of higher vertebrates, each such motor neuron conveys unitary nerve impulses from the spinal cord or brain to a group of striated muscle cells. Such a motor neuron plus the muscle fibers it innervates are called a *motor unit.* The electrical signals traveling along a nerve may reach the *neuromuscular junctions* at rates up to a few hundred per second. A single nerve impulse, or a short volley of impulses, may cause the muscle fibers to give a brief contraction, known as a *twitch contraction.* If the stimulating nerve impulses arrive at a high rate, such that the muscle cannot relax completely between successive stimuli, a stronger, sustained contraction, known as *tetanic contraction,* results. Striated muscle fibers are well adapted to contract and relax quickly. Of course, animals do not move by a series of twitches and convulsions. Skeletal muscles normally contract smoothly and only partially, because not all of the motor units are active at a particular moment. The nervous system controls the strength of the contraction of a particular muscle by controlling the number of motor units that are active. By steady, partial contraction caused by the activity of only a few motor units at any moment, many sets of opposed muscles maintain a steady tonic contraction to hold the animal in any position other than complete relaxation. In these cases, inhibition occurs in the central nervous system, not at the neuromuscular junction.

Not all muscles are controlled in this fashion. In many invertebrates, and in some vertebrate muscles, the motor units associated with a single neuron are larger, and may include an entire muscle. In these cases, the strength of the muscle contraction is not controlled by selecting the number of motor units, but instead the contraction of the muscle can be varied greatly, depending on the frequency at which nerve impulses reach the muscle. In these cases, there are frequently both stimulatory and inhibitory neurons acting on the muscle, and the strength of contraction depends upon their relative effects on the muscle membrane. (See Fig. 6-5.)

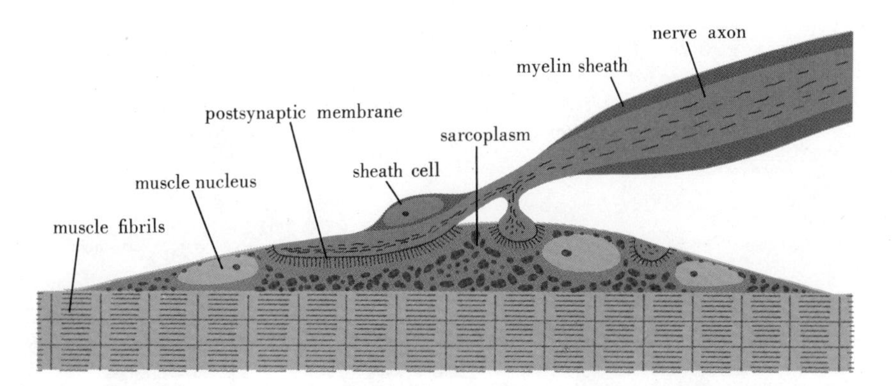


***Fig. 6-5*** *A diagram of a motor end plate in the skeletal muscle of a lizard. The terminations of the nerve lie in synaptic gutters. (After Couteaux, R.,* Exp. Cell Res., *Suppl. 5, 1958.)*

## THE NEUROMUSCULAR JUNCTION

The terminal axon branches of a neuron innervating a muscle fiber have no myelin sheath and come into intimate contact with a specialized portion of the muscle membrane. The region of junction of the muscle fiber and its nerve is called the *end plate.* (See Fig. 6-6.) By use of the electron microscope, the anatomical discontinuity between the nerve fiber and the muscle fiber can be conclusively demonstrated. The terminal nerve endings lie partially enveloped in a trough formed by folds in the muscle cell membrane. (See Fig. 6-7.) These folds in the muscle membrane extend beyond the end plate. A space of about

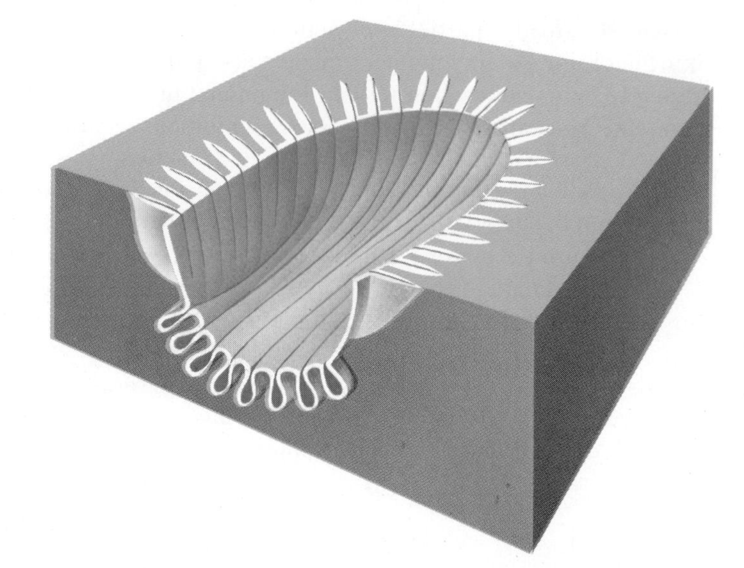

*Fig. 6-6 A diagram of a synaptic gutter or trough formed by the postsynaptic membrane of a motor end plate of lizard skeletal muscle. Acetylcholine is presumably secreted by the axon endings into the trough. The deep folds contain cholinesterase. (After Couteaux, R.,* Exp. Cell Res., *Suppl. 5, 1958.)*

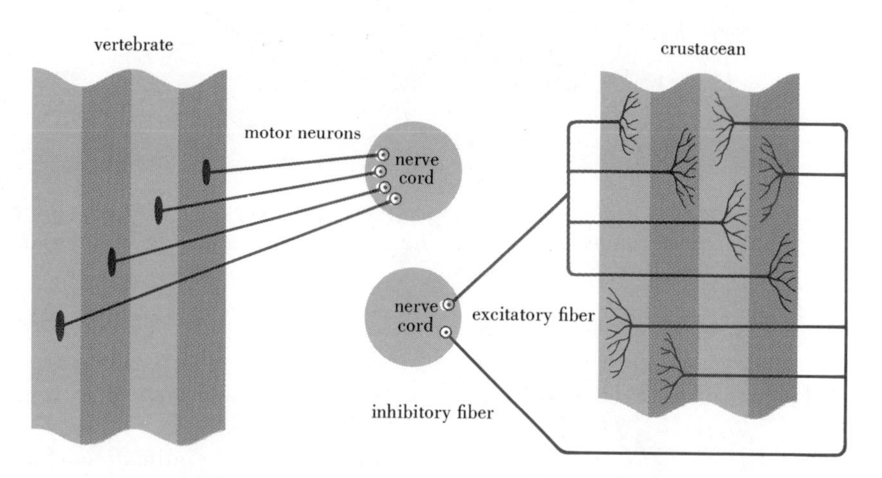


*Fig. 6-7 A diagram comparing the innervation of skeletal muscle in vertebrates with that of the muscles of the appendages in crustaceans. Each motor neuron shown for the vertebrate actually represents a group of neurons forming a motor unit. In addition, a muscle fiber may occasionally have two or three end plates. In the crustacean, only one excitatory fiber is shown but there may be more. Each axon need not branch to each muscle fiber (After W. Hoar,* General and Comparative Physiology, *Prentice-Hall, 1966.)*

several hundred angstrom units separates the axon ending from the muscle membrane as well as one surface of the junctional fold from the other. (An angstrom unit, abbreviated Å, is $10^{-8}$ centimeters.) The nerve endings contain *synaptic vesicles*, bodies of about 300–400 Å in diameter, which are believed to contain the transmitter substance released by the nerve impulse into the end-plate space. In the case of parasympathetic nerves in vertebrates, the substance released is usually *acetylcholine*. Its release into the end plate results in a change of potential across the muscle membrane at that point and may also result in a propagated impulse and a contraction depending upon the number and strength of such stimuli, their timing and spacing, and the state of the muscle fiber. (See Chapter 10.) The released acetylcholine is broken down locally by the enzyme *cholinesterase*, which is present in the end plate. The system has been the subject of a great deal of experimental study. Various drugs may mimic the effect of acetylcholine (parasympathomimetic drugs), block the action of cholinesterase (eserine), or block the action of acetylcholine (atropine). The latter in particular (sometimes in a naturally occurring form called belladonna) is also used in the treatment of peptic ulcers, for example, where the parasympathetic nervous system is involved, or to cause dilation of the pupil of the eye to facilitate eye examination.

Another group of vertebrate neurons, the sympathetic system, usually utilizes *noradrenalin* as the chemical transmitter at the end plate. Those neurons which release acetylcholine are called *cholinergic;* those releasing nor-adrenalin are called *adrenergic*. The effects of locally released nor-adrenalin may be mimicked by systemic adrenalin released by the *adrenal medulla* into the circulatory system. (See Chapter 10.) Further details on the neuromuscular junction and on muscle contraction may be found in *Cell Structure and Function* by Loewy and Siekevitz. (See Further Reading.)

## *LOCOMOTION*

Locomotion is, of course, based upon the use of contractile organs or organelles, usually muscles and cilia or flagella. Similarly, muscles and/or cilia are often used for creating internal or localized external currents in order, for example, to move water over the surface of gills or through a filter-feeding apparatus, to move food along an intestinal tract, or to move blood through a circulatory system.

Microorganisms and small stages in the life cycles of larger organisms (such as larvae and spermatozoa) often move by ciliary or flagellar power. Naturally, such small organisms are moved passively, as well, by currents in the medium in which they live. Such passive movement is often exploited for dispersal to new niches or for dispersion of the young away from competition with their parents. Microorganisms may also be widely dispersed passively by the wind or on the bodies of such freely mobile, fortuitous neighbors as birds. In fact, air currents are also exploited for mobility by larger organisms themselves. In many species of spiders, the young, at a particular point in their development, may gather on exposed surfaces such as the tips of vegetation and spin streamers of silk, which are then caught up by the wind, carrying the young spider to new vistas. Even birds—some albatrosses and vultures, for example—may soar, exploiting the air currents to support their bodies.

With the development of muscle in the coelenterates and other early phyla of animals, active motion by muscular contractions became a possibility. Lacking the rigid skeletal elements that were later to evolve, the coelenterates, flatworms, roundworms, annelids, and echinoderms all developed forms of locomotion based on hydrostatic skeletons. Here, basically, water (or blood) is encapsulated, often but not necessarily in body cavities, and is used to provide a skeleton against which the muscles can act. Water is incompressible, will transmit pressure in all directions, and has low viscosity, allowing ready deformation. With suitable arrangements of layers of muscle fibers—usually one circular and the other longitudinal—imprisoned water can be used to elongate parts of the body while other parts may shorten and thicken. Quite complex repertories of locomotion can be based on such a simple design.

At best, however, hydrostatic skeletons are limited in scope and wasteful of energy. The introduction of rigid skeletons makes possible precision and isolation of movement by permitting a more detailed and precise organization of muscles and by allowing quite specific pairs of antagonists to act together. Leverage is also possible with jointed skeletons so that small contractions can bring about large and powerful movements.

In the arthropods, the appendages form a series associated with the series of segments. Within such a series, regional specialization and division of labor has evolved. (See Fig. 3-11.) In relatively primitive groups, there may be a large number of similar legs. Frequently, however, the anterior appendages are specialized for sensation and for the manipulation of food, and the posterior ones for swimming or walking. The appendages are also often adapted as gills, as parts of a filter-feeding apparatus, or as reproductive accessories.

Flight, of course, has developed *de novo* in insects, in birds, in bats, and in the extinct reptiles, the pterosaurs. In insects, wings are a completely new invention although, of course, they operate by muscle power. In the flying vertebrates, the forelimbs have been borrowed and adapted as wings. The use of the forelegs as wings often restricts their usefulness for locomotion other than flight or for the manipulation of food or for scratching. One can observe substantial rearrangements of body architecture related to flight and to bipedal posture in birds and to flight and suspended posture in many bats. The flexibility (and the length) of most of the vertebral column is usually reduced. The skeleton and musculature of the hind legs and the pelvic girdle are also modified. In birds, the neck is often elongated and ultraflexible to permit the use of the beak as a manipulating organ. In bats, the tail may be greatly elongated and enclosed in a membrane, extending between the legs, that serves as a surface against which the bat may press its prey (in flight) while readjusting its tooth-hold. In some advanced families of bats, walking locomotion, using the wrists as "feet," has re-evolved. Vampires, for example, can walk, jump, or fly with practically equal skill. No birds use their wings for walking. Birds evolved from an ancestral group that was already bipedal (like many of the carnivorous dinosaurs). Although some related lines—crocodilians, for example—have returned to a quadrupedal posture, birds have not.

Another kind of movement, much more subtle than those just listed, exploits the relative density of the animal and the aqueous medium in which it lives. Some fish may change their density by secreting gas into a structure called the *swim bladder*. In the cuttlefish (a mollusk) a special organ, the *cuttlebone*, contains a partially fluid-filled, partially gas-filled chamber, surrounded by a thick connective tissue sheath and by an ion-pumping membrane. The cuttlefish can alter its buoyancy rapidly by pumping ions out of the cuttlebone (water follows passively). Altered buoyancy, of course, means that the animal moves to a new depth in the water.

## ELECTRIC ORGANS

Seven families of fish are now know to possess *electric organs*, derived evolutionarily from muscle fibers. (See Fig. 6-8.) Two of these groups are cartilaginous fish, including the torpedos and some other rays. The marine bony fish, *Astroscopus*, or stargazer, is another example. All other known electric fish are freshwater bony fish—the African electric catfish, *Malopterurus;* the African fish *Gymnarchus;* the family of African "elephant noses," the Mormyridae; and the South American Gymnotidae, including the knife fish and their giant relative, the electric eel.

Since these groups are only remotely related, electric organs appear to have evolved several times independently. In each case, however, they appear to have evolved from muscle. The individual units of such electric organs are very large, syncytial, multinucleate cells lack-

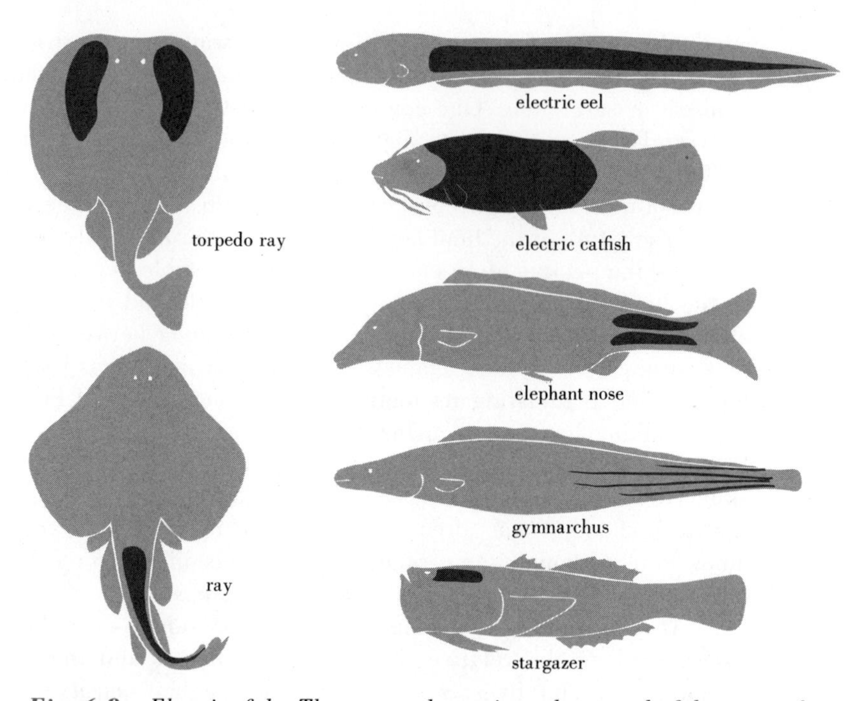


***Fig. 6-8*** *Electric fish. The areas shown in red on each fish are surface projections of the internal electric organs. The organs, of course, are three dimensional. In fact, in some species the electric organs are said to take up over 80 percent of body volume. Each of the fish shown represents an apparent independent development of an electric system.*

ing the oriented contractile fibers of muscle but retaining the impulse generating and/or conducting membrane of muscle and retaining the typical ionic imbalance of nerve and muscle (described in Chapter 10 and in *Cell Structure and Function* by Loewy and Siekevitz) which leads to a resting potential across the cell membrane of about 85 millivolts. The electric units are called *electroplaxes* or *electroplates* and are usually arranged in columns like a stack of coins. That is, most electroplaxes are disk-shaped. (See Fig. 6-9.) Such columns are surrounded by gelatinous connective tissue and then by a sturdy connective sheath and associated with several other similar columns, side by side. In the electric eel, for example, there are from 6000 to 10,000 units per column and some 60 columns on each side of the fish's body. Taken together, the electric tissue in this fish makes up about 40 percent of its bulk. Each electroplax has a relatively smooth, innervated surface and a much-folded opposite surface. All of the innervated faces in a given organ are oriented in the same direction. When such an electroplax is stimulated by its nerve, the polarization of the cell on the innervated face is reversed, the outside becoming about 65 mv negative with respect to the inside. Thus an action potential of about 150 mv results. When all of the electroplaxes, oriented similarly in a column, are stimulated simultaneously, they act like batteries in series; that is, the voltages are summed. If 6000 electroplaxes, each generating 150 mv, are summed, they achieve a potential of 900 volts. Actual measurements on the electric eel have shown levels of 600 volts with a maximum power output of some 100 watts.

In other species of electric fish, the electroplaxes are arranged in much shorter but more abundant columns. The effect here is that of a few batteries in series, but many such short series of batteries arranged in parallel so that only relatively low voltage but high current results. The electric eel, the electric catfish, and the torpedo all release high-voltage discharges. The stargazers and the electric rays produce an intermediate level of several volts, whereas the knife fish, the mormyrids, and *Gymnarchus* emit only about 120 mv. Indeed, the electric eel, as well, has a second set of electric organs that, like those of the knife fish, emit only low voltage.

The functions of these electric organs are only now coming to be understood. The electric catfish seems to discharge only on being touched or alarmed and often follows such action by escape movements. Its high voltage seems adequate to deter potential predators. The electric organ in this fish is wrapped around the body, like a blanket, under the skin.

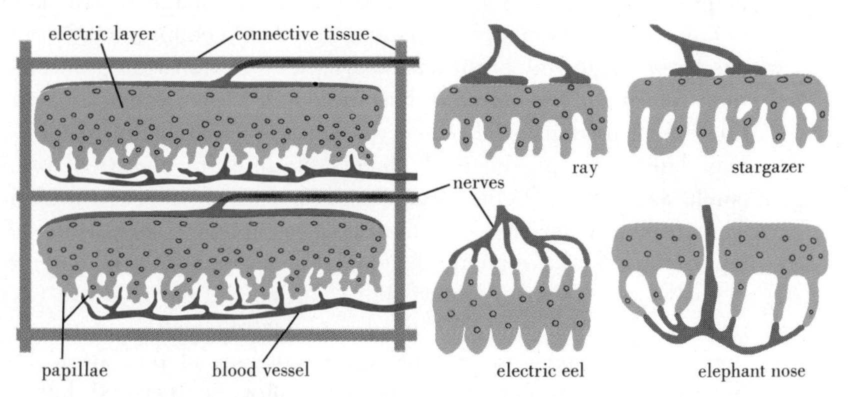


***Fig. 6-9*** *To the left is shown a diagram of the arrangement of the modified muscle cells or electroplates in an electric organ. The cells are organized in stacks, each stack shaped by and each cell contained in a connective tissue compartment. The units of each stack of electroplates are always oriented in the same direction, the broad, innervated faces one way, the papillated faces receiving a capillary blood supply the other way. To the left are shown diagrams of the pattern of innervation of electroplates in four unrelated electric fish. (After Ihle, Dahlgren, and Kepner.)*

The stargazers are marine fish, which often wait quietly, substantially buried in the sandy or muddy bottom, for potential prey to pass by. The eyes are placed in the top of the head, looking upward. The electric organs are derived from the muscles that in other vertebrates move the eyeballs. The very large mouth is held open, often with an attractive, lurelike tongue exposed. When a small fish approaches, apparently attracted by the tongue, the stargazer rapidly engulfs it. The electric discharge occurs at this very moment. We would interpret the function of the discharge to be that of disabling the prey's central nervous system at this crucial moment—either to inhibit escape or to reduce defensive struggling, or perhaps both.

The electric eel uses its high-voltage discharge to stun prey before swallowing them and to shock potential predators. In addition, the electric eel is one of the heroes of comparative biology for its large electroplaxes have served for several decades as a particularly favorable material for the study of the neuromuscular junction, the muscle membrane, and the muscle action potential.

*Gymnarchus*, the mormyrids, and the knife fish are all freshwater species producing trains of low-voltage pulses. The pulses may be emitted only periodically in some species, as when the fish is moving or has been stimulated by the presence of food. In other species, the train of electric pulses may be fixed at a rate of 50 or 500 or even 1000 pulses per second for the animal's entire life, exhibiting only small variations with age and with temperature. Obviously a great deal of metabolic energy is expended by such a system, but the voltage is too low to stun prey or to inhibit predators. A possible communication function has been suggested, but it has not yet been demonstrated in the laboratory. Work on *Gymnarchus* has indicated that the electric organs in these fish are part of a sensory orientation system by which the fish can detect the presence of objects of different electrical properties (such as metals, living organisms, and so forth) in their vicinity. They probably use this system in part to replace vision. Many of these low voltage fish have poorly developed or degenerate eyes and most live in turbid waters with poor visibility during the day and are nocturnal feeders. The sense organs used for *electrodetection* in this orientation system appear to be specialized portions of the lateral line system. Many intriguing questions have now been asked but not yet completely answered about these electric organs and electrodetection systems. How are the discharges coordinated in time so as to give the proper summation of voltage even though the electroplaxes lie at different distances from the central nervous system or from some presumptive initiating center? How do the high-voltage fish protect their own central nervous system from their own discharge? How can they (and perhaps the low-voltage forms, too) keep from stimulating all or many of their sensory nerves by these discharges? By what intermediate steps could such a system have evolved, and is it reasonable for partially parallel cases to have evolved seven times? Some partial answers to these questions and speculations on others can now be found in textbooks of comparative physiology. Other questions remain for future workers.

**OTHER EFFECTORS** Muscle and modified muscle in the form of electric organs are not the only *effectors*, or organs that *do* things. *Glands* are also effectors. Many are stimulated or inhibited by circulating hormones. Such relationships are discussed in Chapter 10. Many glands, however, are innervated by the nervous system much like muscle but not yet as well studied. In mammals, for example, such glands include sweat glands; a variety of dermal, odor-producing glands; salivary glands; glands in the wall of the gut and its derivatives; and glands associated with the reproductive tracts. These may be stimulated and inhibited reciprocally by a double innervation, or the nature of the secretion may be altered by nerve impulses of one or more types. (See Chapter 10.)

Many animals *luminesce*. In many such cases, luminescence seems to serve as a species and/or sex-recognition signal or to coordinate courtship and mating; but other functions have been suggested and, in some forms, no function has yet even been surmised. Some animals luminesce with their own tissues, but many incorporate in or on their bodies, symbiotically, microrganisms that luminesce. In some fish, such symbionts may even be localized in an organ that looks very much like an eye. Often a reflective tissue layer lies behind the luminescent aggregate and reflects the light outward. In some cases, there may even be a movable, opaque lid with which the organ can be covered. But here, in effect, the microorganisms luminesce and the host, at best, simply covers or uncovers them. Some microorganisms themselves, however, and some macroorganisms such as fireflies luminesce periodically or cyclically or on command of their own nervous system.

*Chromatophores* are another widespread type of effector. These are pigment-containing cells, usually located in the skin, which expand or contract under nervous or hormonal control to change the color of the organism. Such color changes may be rhythmic with the day-night cycle—as in the case of many crabs, for example—or may be influenced by the background color, as is true of chameleons and some fish, or they may simply be related to light intensity. Usually such changes decrease the visibility of the animals to predators. In other cases, such changes may be used to increase or decrease heat absorption. Changes in chromatophores with emotion are also common, especially in fish, where fright or courtship, for example, may lead to changes in color pattern that make the fish almost unrecognizable as the same species.

## *FURTHER READING*

Bloom, W., and D. W. Fawcett, *A Textbook of Histology*, 9th ed. Philadelphia: Saunders, 1968.

Bourne, G. H. (ed.), *The Structure and Function of Muscle.* Vols. 1–3. New York: Academic Press, 1960.

Cahn, P. H. (ed.), *Lateral Line Detectors.* Bloomington: Indiana University Press, 1967.

Davson, H., *A Textbook of General Physiology*, 3d ed. Boston: Little, Brown, 1964.

Florey, E., *An Introduction to General and Comparative Animal Physiology.* Philadelphia: Saunders, 1966.

Hoar, W., *General and Comparative Physiology.* Englewood Cliffs, N.J.: Prentice-Hall, 1966.

Hoyle, G., "Neuromuscular Physiology," *Advances in Comparative Physiology and Biochemistry*, 1: 177-216, 1962.

Katz, B., *Nerve, Muscle and Synapse.* New York: McGraw-Hill, 1966.

Laverack, M. S., *The Physiology of Earthworms.* New York: Macmillan, 1963.

Loewy, A. G., and P. Siekevitz, *Cell Structure and Function*, 2d ed. New York: Holt, Rinehart and Winston, 1969.

Romer, A. S., *The Vertebrate Body*, 3d ed. Philadelphia: Saunders, 1962.

Smith, D. S., "The Flight Muscles of Insects," *Scientific American*, 212 (6): 76–88, 1965.

chapter 7

# Circulatory Systems

The importance of internal circulation for an animal of any considerable size has already been emphasized in Chapters 2 and 5. Protoplasmic streaming provides a sort of circulation within many single cells; in coelenterates the contractions of the body wall redistribute the water contained within the digestive cavity. The need for oxygen in order to make efficient use of food molecules has led to the presence in most of the more highly organized animals not only of respiratory organs but of efficient arrangements for respiratory gas transport via the blood. The capacity of the circulatory system often limits the activities and effectiveness of larger animals. Failure of the circulatory system from hemorrhage or loss of pumping capacity is a common cause of death in human beings. The price of large size and an active life is a basic dependence on our circulatory system.

The fluid flowing in the circulatory systems of animals is basically a salt solution not greatly different from the fluid portion of protoplasm in the proportions of the various ions, sometimes with the exception of $Na^+$ and $K^+$. Macromolecules such as proteins are dissolved in the blood, and a variety of blood cells are usually also carried. In vertebrates, the most abundant of these cells are the *red blood cells*, or *erythrocytes*, which are substantially filled with a reddish protein called *hemoglobin*. The hemoglobins of various animals differ slightly; the amino acid components vary in number and sequence. Some of these differences may be related to the animal's way of life. Blood also contains amoeboid cells of various types, called *white blood cells* or *leucocytes*. In vertebrates, there are also fragments of cells, called *platelets*, that are active in the blood-clotting mechanism. Among the many proteins carried in the fluid portions of the blood, some help to initiate the formation of blood clots when exposed to air or when they pass through a break in a blood vessel. Others, called *antibodies*, combine with foreign proteins or other large molecules and thus participate in their removal from circulation; still others are *hormones* (not all hormones are proteins but all are carried by the blood) serving as chemical messengers coordinating the functions of different parts of the animal's body. There may also be parasitic organisms living in the blood. Although one of the primary functions of the blood in most animals is to aid in respiratory gas exchange, many other substances such as fat droplets and food molecules are also transported and exchanged, via the blood, between different parts of the body. Blood also serves as an important link in the systems regulating water and ion balance and in the delivery of water and ions to the proper sites, for the removal of metabolic products from cells, and for the distribution, in some animals, of heat throughout the body.

## THE STRUCTURE OF CIRCULATORY SYSTEMS

Some annelid worms have excellent circulatory systems that display all the major features found in more elaborate animals; others have much simpler ones or none at all. Hence these animals provide examples of all major features of circulatory systems. A striking difference between the circulatory system shown in Fig. 7-1 and that of most vertebrates is the presence of several *hearts* or muscular thickenings of the walls of the larger blood vessels. Many of the larger vessels undergo peristaltic waves of contraction, much like those that characterize intestinal tracts.

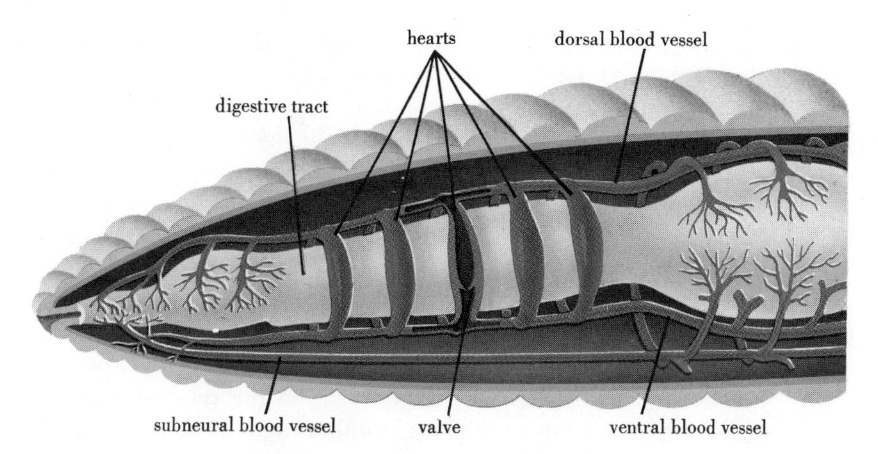


***Fig. 7-1*** *Part of the circulatory system of an annelid. Capillaries and sinuses cannot be shown at this scale.*

The larger vessels with muscular walls are called *arteries*; in annelids there is often no sharp division between the thicker arteries and the several hearts. The larger arteries branch, usually dichotomously, into successively smaller ones with thinner or less muscular walls. The limit of this latter process is reached in a type of blood vessel called a *capillary*, which has walls only one cell thick; often these cells of the capillary wall are flattened so that less than a micron of protoplasm separates the blood from the tissue outside of the capillary. Most small molecules diffuse easily through the walls of capillaries; amoeboid cells circulating in the blood often push their way between the cells forming capillary walls and thus may enter or leave the circulatory system proper.

Individual capillaries are usually a fraction of a millimeter in length but, taken together, have an enormous surface area. Capillaries join one another and connect to larger vessels called *veins*, which become larger as they proceed and receive blood from other small veins or *venules*. The largest veins connect, ultimately, with the arteries and heart; hence the blood flows around the circuit or, rather, may flow around any of a number of circuits, depending upon which branch a given element of blood takes at each of the countless branchings of the arteries and *arterioles*.

In addition to this type of *closed circulation*, in which blood is carried from the heart around through capillaries and back to the same or a different heart, some arteries of some annelids discharge blood into *sinuses*, irregular spaces between cells, rather than into closed capillaries. There are also some veins that have openings by which blood re-enters the circulatory system from the sinuses. In one animal, there may be both closed and *open circulation* (in which blood follows no clearly defined pathway for part of its journey). Where one type greatly predominates, the circulatory system as a whole is called by that name; for example, arthropods and most mollusks have an open circulatory system, whereas vertebrates have a closed one.

The strongest muscular heart will not move blood around a closed circulatory system unless there are *check valves*, which permit forward flow but prevent back flow. All hearts are provided with effective valves without which they would be useless. Less obvious, but almost equally important for an effective circulation, are valves located in the veins. (See Fig. 7-2.) In the veins, the pumping action of the heart or hearts exerts little hydrostatic pressure owing to the relatively large volume of the capillary beds or sinuses and their high friction and lack of tone. Contractions of muscles in the body wall, however, may serve to move blood back toward the heart even though these muscles are not part of the circulatory system proper. By such intermittent muscular squeezings of the veins, blood is moved in the direction permitted by the check valves. In this way blood is also moved out of open parts of a circulatory system. (In vertebrates, the tone of the smooth muscles in the walls of veins varies with the animal's needs. The veins then serve as a blood volume reservoir.) Circulation in an open system is less rapid than in a closed one but more effective in one sense because blood can flow in immediate contact with cells rather than being separated from them even by the thin wall of a capillary. In the dynamically closed system, on the other hand, the ubiquitous capillaries reach the near vicinity of more cells than are reached by the free-flowing blood of open systems. Another important difference between these two designs is that open systems seem to lend themselves to use as hydrostatic skeletons. Closed circulatory systems may also serve as hydrostatic skeletons, however,

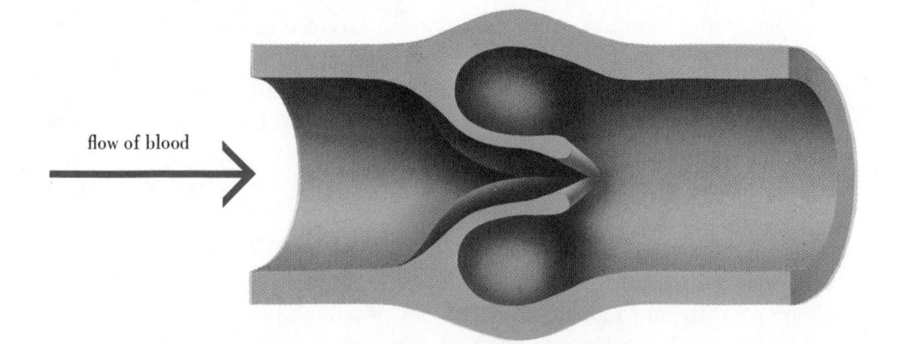


***Fig. 7-2*** *Valves such as this one, in the veins of mammals, aid in the return of blood to the heart against gravity.*

as in the case of erectile tissue in the penis of many mammals. The mechanisms are, of course, quite different.

In mollusks and arthropods, the circulatory system is usually less elaborate than in many of the annelid worms. (In squids, the circulatory system is of advanced design.) From the heart, a few arteries lead to major organs of the body, but large parts of the course followed by the blood lie outside of definite blood vessels, and thus these circulatory systems are open ones. In a squid, there is a separate heart to pump blood through the gills. In many arthropods, there are no veins leading directly into the heart; instead, the heart is located in a large blood sinus that is fed by other sinuses leading from all parts of the animal's body. Blood re-enters the heart through lateral openings, called *ostia*, provided with check valves. As the heart contracts, the valves close and blood is forced forward into the arteries. On relaxation of the heart, the natural elasticity of the muscle and connective tissue in its walls and external suspensory ligaments cause the internal volume to increase; the valves then open and blood flows in from the surrounding blood sinus. There are also valves at the entrance to the arteries to maintain the unidirectional flow of blood. Many of the sinuses, furthermore, are fairly definite channels that lead from one diffuse space to another; thus, in effect, there are venous channels in the body to aid in the return of blood to the sinus surrounding the heart.

## *FUNCTIONAL ADAPTIONS IN THE CIRCULATORY SYSTEMS OF VERTEBRATE ANIMALS*

In vertebrates, with a few exceptions, to be noted, the blood is retained in distinct vessels. More important, however, is the high degree of specialization of the various parts of the system and the efficiency of transportation of blood to every part of the animal's body. This specialization has been necessitated by the large size and high degree of activity of most vertebrates. Our major rivals in activity are the insects, in which the tracheal system replaces the blood for respiratory gas transport and which, therefore, need no elaborate circulatory apparatus. The maximum hydrostatic pressure produced by the pumping action of the heart provides an index of the effectiveness of the circulatory system in rapid transport of blood. Since this pressure is most easily measured in the larger arteries, it is usually known as the *arterial pressure*. In popular and medical usage, it is often called simply "blood pressure," a term that on closer examination proves quite misleading since the pressure of the blood varies widely in different parts of the circulatory system. In selected animals of various major phyla, maximum arterial pressure at the peak of cardiac contraction has the following approximate values when the animal is at rest (during exercise the arterial pressure rises somewhat but seldom does it as much as double): annelids and arthropods, 5–10 mm Hg (open system); active fish such as salmon, 75 mm Hg in the ventral aorta before the gills and 50 mm Hg in the dorsal aorta after the gills have been passed; birds and mammals, 120–180 mm Hg. The higher arterial pressure of birds and mammals varies rather little between large and small species. The giraffe, a very interesting exception, requires a high arterial pressure for blood to reach the head. The pressure in an open system cannot, of course, be directly compared with the pressure in a closed system.

An important aspect of vertebrate circulatory systems is the elasticity of the walls of the arteries, produced by the intrinsic elasticity of the encircling smooth muscle cells and by the presence of a large amount of elastic connective tissue. The high arterial pressure resulting from cardiac contraction stretches the arterial walls much as rubber tubing can be stretched by water pressure. This stretching, in turn, creates a small reservoir of blood under moderate pressure, which, during the relaxation phase of the heart, will continue to be forced through the smaller, branching arteries and the capillaries. Thus the arterial pressure rises sharply at each heartbeat to cause the pulse; it also remains quite high throughout the cardiac cycle because of this arterial elasticity and the resistance to flow created by the friction of the capillary bed. The walls of the smallest terminal arteries, called arterioles, have relatively more smooth muscle than the larger arteries. These muscle cells may contract, under the control of the sympathetic nervous system or the adrenal medulla, to reduce the flow of blood through arterioles. This is the basic mechanism by which the relative amount of blood reaching various organs can be controlled. Not only are the arterioles independently controlled by the nervous system but their muscle cells apparently respond differently to identical or similar hormonal stimulation; that is, some arteriolar muscles contract and others do not. Such differential responsiveness leads to shifts in blood volume and flow from one organ to another.

The small blood vessels, arterioles and capillaries, taken together, have a high inside surface area compared with the large arteries, taken together. In many capillaries, the lumen is so narrow that the blood cells must actually be deformed to get through. Blood flow is thereby retarded by friction, and most of the arterial pressure is lost, with the result that the heart-derived hydrostatic pressure remaining by the time the blood reaches the veins is very small. In contrast with the arteries, veins have relatively thin walls, with few smooth muscle cells and less elastic connective tissue. They are, however, sufficiently muscular to expand or contract under sympathetic nervous control and thus increase or decrease the pooling of blood or its return to the heart. They serve as a blood reservoir.

The dissipation of arterial pressure as blood passes through capillaries creates a practical problem; that is, how is the blood to complete the rest of its circuit back to the heart? In most cases, the veins draining blood from a given organ are approximately as long as the arteries supplying it, yet there is no large pressure difference to force the blood on its way. Large, terrestrial mammals potentially have an added problem in that gravity retards the return of venous blood from part of the body lying below the heart. This problem is overcome by the widespread presence in veins of check valves that prevent backflow. (See Fig. 7-2.) The forward propulsion of blood toward the heart, just as in the circulatory systems of smaller and simpler animals, is largely caused by the contractions of surrounding skeletal muscles. In mammals, the heart is contained in a connective tissue sac, the *pericardium* and is separated from the pleural cavity containing the lungs. But the pericardium is flexible enough that expansion of the chest to fill the lungs serves also to draw venous blood into the heart.

The fact that venous blood returns to the heart at a low pressure, dependent in part on muscles that are not really part of the circulatory system at all, means that filling of the heart requires special engineering measures. In all vertebrates, and also in a few of the more active invertebrate animals such as squids, the heart consists of at least two chambers arranged in series; these chambers are usually called the *atrium* and *ventricle*. (See Fig. 7-3.) In mammals, for example, the atrium

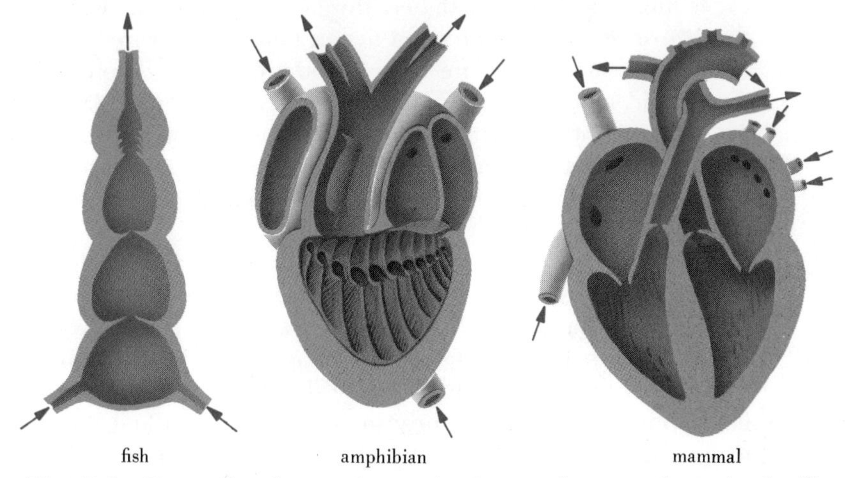


***Fig. 7-3*** *Heart chambers and organization at three vertebrate levels. The first is a diagrammatic representation of the heart of a primitive fish. Blood enters the relatively large, but thin-walled, sinus venosus and passes thence into successively thicker-walled chambers with smaller lumens. Amphibians are represented here by a frog in which the atrium is divided into two by a septum. The "single", complexly divided ventricle leads to a peculiarly subdivided aorta. Physiologically the frog achieves a double circulation. Mammals have a complete, anatomically double circulation.*

receives blood from the larger veins and is relatively thin-walled. It contracts shortly before the ventricle does and fills the ventricle with blood under moderate pressure. The ventricle, with much thicker walls and a smaller lumen than the atrium, contracts against this charge of blood and its more powerful musculature raises the pressure to the arterial level. There are check valves between atrium and ventricle and between the latter and the *aorta* (the first, main artery in mammals). Having two such chambers arranged in series has a double advantage in that the pumping is divided between two masses of muscle and the preliminary contraction of the atrium distends the ventricle by forcing blood into it and stretches its muscle fibers so that on contraction they can develop more force. This greater force results, in part, from the general property that muscles have of being elastic as well as contractile.

## ORGANIZATION OF VERTEBRATE CIRCULATORY SYSTEMS

In fish, the heart is usually more complicated, linearly, than in mammals. There is an initial, additional, large, very thin-walled chamber, receiving venous blood, called the *sinus venosus*. When the heart contracts, the blood is passed into the atrium, thence into the ventricle, and finally into a very narrow and exceedingly muscular chamber, the *bulbus arteriosus*, which leads to the *ventral aorta*. Here, then, a linear, four-chamber system is utilized to raise the blood pressure by sequentially reducing the caliber of the lumen and increasing the muscular power of the wall. (See Fig. 7-3.) The ventral aorta leads forward from the heart and terminates as four or more pairs of *afferent arteries* to the gills. These arteries, arranged symmetrically and bilaterally, pass into the gill region and there branch into arterioles and finally into capillaries.

The gill capillaries run just beneath the surface of the gill tissue picking up oxygen and also often making some exchange of ions, carbon dioxide, ammonia, or other solutes with the passing water. These gill capillaries unite to form *efferent arterioles* and arteries (usually four pairs), which join and form the *dorsal aorta*; the latter branches both anteriorly and posteriorly to supply the head and the body with oxygenated blood. The dorsal aorta and its major branches divide and subdivide to supply arterial blood to the several organs of the body.

Most of the blood supplying the digestive tract, after passing through the capillary system, picking up sugars, amino acids, and other small molecules produced by digestion, is regathered into a large vessel, the *hepatic portal vein*, which instead of leading back to the heart delivers the deoxygenated but food-charged blood to the liver.

There the vein subdivides into capillarylike vessels called *sinusoids*. These sinusoids, in turn, regroup into veins, which lead back to the heart. Most of the blood to the posterior part of the body—especially the pelvic region and the tail—after passing through the initial capillary system in the tissues is regrouped into the *renal portal veins*, which deliver the deoxygenated blood, charged with metabolic waste products, to the kidneys. Here the renal portal veins subdivide eventually into the *peritubular capillaries*. (See Chapter 8.) These capillaries regroup into the *renal veins*, which lead, eventually, back to the heart. Both organs supplied with deoxygenated blood—the liver and the kidneys—also receive a "normal", arterial blood supply via arteries branching directly from the dorsal aorta. Note that the portal blood in the liver sinusoids or in the renal peritubular capillaries passes through three capillary systems in making a full circuit. All of the blood in a fish must first go through the gill capillaries and through a second set of capillaries in one of the body organs. Only in the case of the blood to the digestive tract or to the posterior part of the body is there a third set of capillaries to be traversed in the liver or kidneys, respectively. (See Fig. 7-4.) At

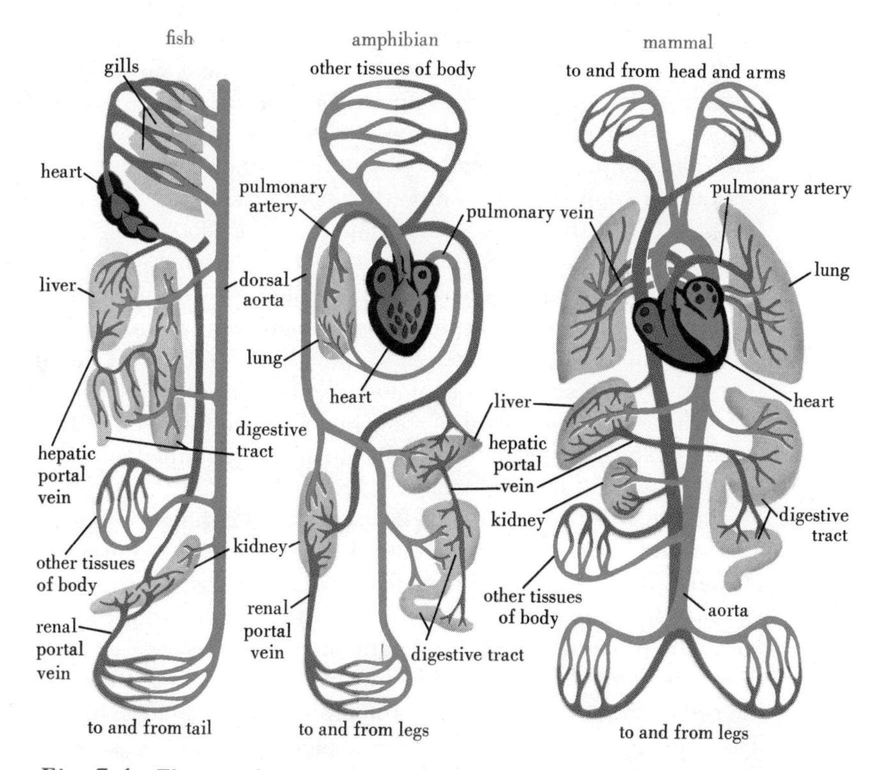


***Fig. 7-4*** *The circulatory system of three types of vertebrates, emphasizing the pulmonary (or gill), systemic, renal portal, and hepatic portal systems.*

each set of capillaries, the blood pressure drops considerably; the flow in the venous systems, including the portal veins, is assisted by valves and by the squeezing action of surrounding musculature and body movements.

In inquiring why the circulatory system should be organized in this way, we must first note the advantage gained by having all blood flow first through the gills for respiratory gas exchange. Wherever it may go next, it will now be well supplied with oxygen. Against this must be weighed the inefficiency caused by loss of some arterial pressure as the blood passes through the gill capillaries. The hepatic portal system serves to carry products of digestion in high concentration directly from the digestive tract to the cells of the liver, where many of them are stored or chemically converted. The advantages of the renal portal system have not yet been completely worked out. A large pool of blood at low pressure, poor in oxygen, but rich in metabolites, is supplied to the renal tubules, where secretion and/or absorption of water, ions, nitrogenous wastes, and other substances, takes place. More we cannot say as yet. In mammals, the renal portal system has been eliminated, but replaced by a new design that may do the same things. The arterial supply to the mammalian kidney subdivides into clumps of *glomerular capillaries* at the proximal end of each renal tubule. The capillaries regroup as *efferent arterioles*, which form a second group of capillaries, the *peritubular capillaries*, surrounding most parts of the renal tubules except the glomerulus. The venous blood from these capillaries goes directly back to the heart. (See Fig. 8-5 and Chapter 8.) In the glomerular capillaries, the blood does not give up oxygen as much as it does fluid volume. Thus, in mammals, oxygenated blood flows through the peritubular capillaries. The peritubular blood is also not particularly enriched with metabolic products. The blood pressure, however, is higher in mammals and the blood flow through the peritubular capillaries is almost surely swifter in mammals than in fish. There being no renal portal system, in mammals, the blood from the capillaries of the hind portion of the body simply returns directly to the heart via "normal" veins.

In terrestrial vertebrates, or tetrapods, lungs usually replace gills as the respiratory organs and the arrangement of the blood circuit changes. Blood from the lungs, freshly oxygenated, returns to the heart via *pulmonary veins* and enters the *left side* of the atrium which is now normally separated from the right side by a *septum* or wall. Blood returning from the body and head, the *systemic circulation*, flows into the *right atrium* via a modified, and often much reduced, sinus venosus. Thus the oxygenated and oxygen-depleted bloods are kept separate. In amphibians, each atrial half opens via its own *atrioventricular canal*, guarded by valves, into a single, multiply subdivided ventricle. The ventricle is so crisscrossed by muscular and connective tissue *trabeculae* as to be almost spongelike and mixture of the two bloodstreams seems not to occur to any great extent. (See Fig. 7-3.) The *conus arteriosus* (as the exit chamber is now called) is also subdivided by a *spiral valve.* On ventricular contraction, the left-sided, oxygenated blood appears to be guided into the *systemic aorta* while the deoxygenated blood from the right side flows into the *pulmonary arteries*, which are new branches of the most posterior pair of the former gill arteries of fish. Thus there is now a *double circulation*: sinus venosus to right atrium to ventricle to pulmonary arteries to lung capillaries to pulmonary veins to left atrium to ventricle to systemic aorta to head and body to systemic capillaries and back directly to the sinus venosus or back via the hepatic or renal portal systems. Amphibians exhibit almost all conceivable exceptions to this diagrammatic scheme. In many, respiration is substantially dermal or pharyngeal. Some have external gills, with or without lungs, for part or all of their life. Some have no lungs or gills whatsoever. Each variant exhibits suitable rearrangements of the blood's circuit. Lungless amphibians, for example, have no pulmonary arteries and have a perforated or vestigial interatrial septum, so no double circulation exists. They are normally inactive animals dependent on dermal respiration. As was the case with other organs (described earlier in this chapter) that receive deoxygenated blood supplies, terrestrial vertebrates with a pulmonary circulation of deoxygenated blood also have an arterial blood supply, branching from the systemic aorta, to the tissue of the lung itself.

In reptiles, a double circulation is universal—as, indeed, is pulmonary respiration (although some turtles supplement pulmonary with dermal, pharyngeal, and cloacal respiration). Reptilian hearts are complex internally; all have complete physiological separation of the two bloodstreams throughout the atrial and ventricular pathways. Functional septal arrangements in turtles, lizards, and snakes are still to be fully worked out. In crocodilians, there are totally anatomically separate atria and ventricles, the "right" side pumping deoxygenated blood to the lungs; the "left" side pumping oxygenated blood to the body. The same is true of birds and mammals. In all three groups, the sinus venosus and conus arteriosus have disappeared as conspicuous anatomical chambers. Thus fish hearts have four chambers in series; living amphibians have four functional chambers on the right side and three functional chambers on the left side of the heart. Reptiles, birds, and mammals can best be said to have two side-by-side, two-chambered series, four chambers in all.

Before we leave this complicated but still greatly simplified view of vertebrate circulatory systems, let us examine another key portal system, the *pituitary* or *hypophyseal portal system.* In mammals (the system varies from group to group among vertebrates but the basic principles seem to hold true), the anterior lobe of the pituitary gland or *adenohypophysis* (which produces at least six crucial hormones, described in Chapter 10) receives no direct arterial blood supply. The only blood reaching this glandular tissue first runs through a capillary bed in the *median eminence* of the *hypothalamus* – a portion of the floor of the forebrain which is immediately adjacent to the pituitary. The capillaries of the median eminence coalesce into efferent arterioles and then flow to the anterior lobe of the pituitary. (See Fig. 10-10.) The system is apparently designed to pick up, in minute amounts, hormonal messages, secreted or released locally by the neurons of the hypothalamus, and to deliver these to their target organ, the anterior pituitary, directly and efficiently. On this hypophyseal portal system depends the integrity of integration of the mammalian body. It is the principal link between the two integrative systems – the endocrine system and the nervous system.

In all vertebrate animals, there is still another part of the circulatory system, the *lymphatic vessels.* These thin-walled vessels begin as finely branched tubes lying between cells almost everywhere in the body excepting only the central nervous system. Lymphatics are almost as small as capillaries but are not connected to arteries. Small lymphatics join to form larger ones; the largest ones drain into veins. Lymphatics are provided with valves. In some amphibians there are small *lymph hearts* that contract rather feebly to pump lymph into the veins. Otherwise, flow is entirely maintained by the squeezing action of surrounding muscles. The *lymph*, the fluid contained in the lymphatics, is colorless owing to the absence of red blood cells but is otherwise similar to the blood. Lymph enters the small terminal lymphatics by diffusion from surrounding intercellular spaces. Indeed, lymph is extracellular fluid that has entered lymphatic vessels. In its extravascular role, the lymph or extracellular fluid bathes the cells of the body. Thus nutrients, oxygen, and hormones diffusing from capillaries cross to cells via the lymph and are part of it, whereas carbon dioxide, urea, lactic acid, and so forth, diffusing out of cells, also pass through the extracellular fluid en route to capillaries. Amoeboid cells (the white blood cells) sometimes force their way between the thin cells forming the walls of the lymphatics, and hence are found in the lymph. Although the lymphatic system is not an open one in structure, since all its vessels are surrounded by distinct walls of cells, its function resembles that of the open blood sinuses of invertebrates in that the lymph can work its way back to the heart from all parts of the body. This process serves to equalize distribution of body fluids. Lymphatics originating in the intestinal villi also carry most of the absorbed fat from the intestine to the systemic veins, thus bypassing the hepatic portal system and the liver.

Associated with the lymphatic system, and also with the circulatory system in general, are certain specialized organs whose complete function is not yet fully understood. This group of organs includes the *lymph nodes*, the *spleen*, the *thymus*, and more or less diffuse patches of lymphoid tissue spread throughout the body but often concentrated under the mucus membrane lining the gut or in the wall of the gut. The lymph nodes, the thymus, and the spleen, as well as other patches of lymphoid tissue—and, in some cases, renal tissue—appear to have the function of producing one of the types of leucocytes, the *lymphocytes*, that are involved in immune reactions, inflammation, and repair. The thymus has also been strongly invoked, in young mammals, as an important organ in the development and maintenance of immune reactions. The spleen produces certain other blood cells, in addition to lymphocytes, and destroys aged red cells as well.

There are also blood sinuses in vertebrates, but they are restricted to a few specific organs. The spleen, for example, contains a spongy mass of sinuses that serve to store blood. In the placenta of pregnant mammals, blood also fills sinuses; the exchange of materials between maternal and fetal blood occurs between such juxtaposed sinuses. (See Fig. 9-3.)

## CHEMICAL SPECIALIZATIONS OF BLOOD

Blood is more than a simple salt solution populated by drifting cells; it contains several types of molecules adapted for the efficient transport of oxygen. Molecular oxygen is not a highly reactive substance and, unlike carbon dioxide, it can only be held in water in physical solution. (See Chapter 5.) When water comes into equilibrium with air, which ordinarily contains just under 21 percent oxygen, it takes into solution only about 0.5 percent of its volume of gaseous oxygen. Expressed in terms of weight, this is only a few parts per million. Hence only a small fraction of the blood pumped through the circulatory system can consist of oxygen in physical solution. Since the rate at which oxygen can be transported is often a limiting factor in the success of the animal, it is clearly advantageous to increase the capacity of the blood to carry oxygen. This is accomplished in vertebrate animals by *hemoglobin*, a reddish, iron-containing protein located inside erythrocytes. Some mollusks and arthropods have, instead, a copper-containing protein called *hemocyanin* in solution in their blood. Many annelid worms also have hemoglobins in their blood, but these are free of blood cells, occurring as large, polymeric molecules directly dissolved in the plasma. Placing hemoglobin inside the red cells or in very large molecules reduces its influence on the viscosity and the osmotic pressure of the blood.

The following typical values of the oxygen carrying capacity of the blood of various animals when in equilibrium with air show the effectiveness of hemoglobin and hemocyanin: mollusks and arthropods employing hemocyanin, 1–4 ml of gaseous oxygen per 100 ml of blood; some of the burrowing marine annelids such as *Arenicola*, 9 ml/100 ml; fishes, 10–16 ml/100 ml; most birds and mammals, 15–20 ml; and marine mammals such as seals and porpoises, which use hemoglobin as a means of storing some oxygen for long dives, up to 30 ml of oxygen per 100 ml of blood.

Hemoglobin, or other substances that serve the same function, not only must take up oxygen at the respiratory surfaces, but also must give it off again in the vicinity of cells that require it for their metabolism. A number of chemical compounds combine with oxygen, but do it so firmly that they would be worse than useless for oxygen transport in an animal's blood, because the combination would be so stable that the animal would asphyxiate even though its blood was loaded with oxygen. What is needed is a molecule with flexible powers of combining loosely with oxygen at the respiratory organs and then releasing it equally freely in capillaries within easy diffusion range of actively metabolizing cells.

To appreciate the biochemical adaptations of blood, one must distinguish between the concentration of oxygen in solution in the blood and the further quantity of oxygen that is present in chemical combination with hemoglobin. Atmospheric oxygen is in equilibrium with the oxygen dissolved in the blood plasma. The oxygen dissolved in the blood plasma is also in equilibrium with the oxygen combined with hemoglobin in the red cells. This latter quantity is, as noted previously, by far the larger amount. Certain marine fishes, entirely lacking red cells and hemoglobin, function on dissolved oxygen alone. The oxygen carrying capacity of their blood is only about 0.9 volumes percent. (Note that their blood is in equilibrium with the same partial pressure of about 150 mm Hg of oxygen in the atmosphere as are avian and mammalian bloods with 15–25 volumes percent carrying capacity.) Part of the compensatory mechanism of these fish is apparently a sluggish way of life.

The concentration of oxygen dissolved in blood plasma will vary directly with the partial pressure of oxygen in the atmosphere. This variation is, of course, important for organisms that live in environments low in oxygen, but is negligible for organisms living in the air or well-aerated waters. The oxygen-carrying capacity of hemoglobin or other blood pigments also varies with the partial pressure of oxygen with which it is in equilibrium. At some partial pressure of oxygen, however, the hemoglobin is finally completely saturated—that is, it can no longer combine with more oxygen. In vertebrate organisms this total saturation is often at about the partial pressure of oxygen in the normal atmosphere. The amount of oxygen being carried by the blood under given circumstances in a given organism is often expressed as percent saturation, where "saturation" refers to the total carrying capacity of the blood at the partial pressure at which the hemoglobin is completely combined with oxygen.

In most natural bodies of water, including the oceans, there are only enough organisms to lower the partial pressure of oxygen slightly below that in the atmosphere (about 21 percent of 760 mm Hg). Hence most animals find oxygen in their environment at a partial pressure of roughly 150 mm Hg. At this partial pressure, their blood takes up oxygen and almost exactly the same partial pressure is retained as the blood flows from the respiratory surface to the vicinity of cells in active tissues. The more active the metabolism of these cells, the lower will be the partial pressure of oxygen in their immediate vicinity, since metabolic processes normally use up oxygen. A concentration gradient is thus set up from the blood to the intercellular fluid to the cells; the oxygen diffuses, therefore, from the blood into the cells, along this concentration gradient.

Hemoglobin must now reverse the role it plays at the respiratory surfaces; the more rapidly and completely it can give off oxygen, the better it will serve the needs of the animal. In response to this need, the hemoglobin of active animals has the property of combining vigorously with oxygen only over a certain range of high partial pressures and releasing oxygen at lower partial pressures. This is illustrated in the graphs shown in Fig. 7-5, 7-6, and 7-7 in which the degree of saturation

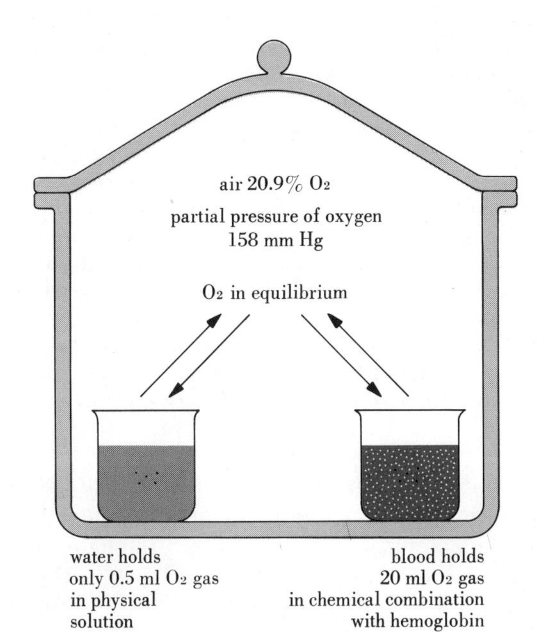


***Fig. 7-5*** *Water and blood, in this case, hold oxygen at the same partial pressure (both are in equilibrium with the same gas), and with about the same concentration of oxygen in solution. In water, however, oxygen is held only in solution; in blood it is held also in chemical combination with hemoglobin.*

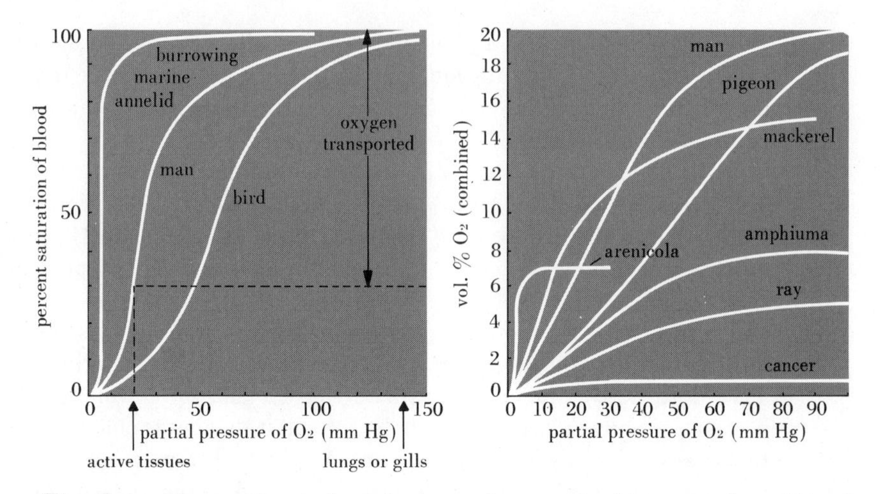


***Fig. 7-6*** *The graph on the left shows the quantitative properties of blood in three types of animals adapted for life in different environments. The burrowing marine annelid,* Arenicola, *lives in very oxygen-poor, muddy sediments. Men and birds both occupy the same environment in terms of high partial pressures of oxygen. Birds, however, have achieved greater efficiency of oxygen delivery, perhaps in association with the heavy metabolic demands of flight. One can read from a graph such as this the percent of oxygen which the blood of a particular species will deliver to the tissues provided one knows the general level of saturation achieved at the respiratory surface and present in the metabolizing tissues of the body. An example is shown for man. The blood normally becomes almost totally $O_2$ saturated in the lungs. Active body tissues often have partial pressures of $O_2$ of about 20 mm Hg. Such partial pressure of $O_2$, in human blood, occurs at 30 percent saturation. Thus totally saturated human blood yields about 70 percent of its oxygen as it flows through capillaries in active tissues. If, in birds, the partial pressure of $O_2$ in active tissues were the same, about 90 percent of the oxygen which had been taken up at the lungs would then be yielded. Bloods of different groups have different oxygen carrying capacities depending upon the presence and form of $O_2$ carrying molecules, temperature, pH, and many other features. In order to permit comparison of these different bloods we have chosen for the ordinate, percent saturation which is the maximum amount of oxygen which a particular blood will hold when in equilibrium with air. At the right we have shown the actual amount of oxygen carried by the blood of several animals at different partial pressures of oxygen and at physiological temperatures and pH.* Arenicola *can be seen to take up $O_2$ at low partial pressures but to a limited extent. The crabs,* Cancer, *the rays and* Amphiuma, *a salamander, lead relatively quiet lives in well oxygenated environments. Advanced, active vertebrates such as many bony fish, birds, and mammals have high oxygen carrying capacities. (Partly after Redfield from Prosser and Browning.)*

of the blood with oxygen is plotted against the partial pressure of oxygen with which it is in equlibrium. Note that in birds and mammals the curve has an S shape. The placement and shape of the S indicate that their hemoglobins will be almost totally saturated with oxygen at the partial pressure present in the lungs, about 150 mm Hg. In birds, all but about

10 percent of this oxygen will be yielded in the capillaries, where the partial pressure of oxygen in the surrounding metabolizing cells is about 20 mm Hg. In mammals, at 20 mm Hg, all but about 30 percent of the oxygen in the blood will be yielded to the cells. The shape of these curves further indicates that birds will be in serious trouble if the partial pressure of oxygen available at the lungs is less than, say, 60 mm Hg. At that level, the blood would only become about 50 percent saturated to begin with and, therefore, would not, at the capillaries, yield enough oxygen to sustain metabolism. For man, with a rather differently shaped curve, achieving the same (40 percent of saturation) yield of oxygen, is possible even if the partial pressure in his lungs is as low as about 32 mm Hg. But note the curve for the burrowing marine annelid. At the partial pressure at which bird and human hemoglobins are expected to yield their oxygen at the cells, the annelid hemoglobin is almost completely tied to its oxygen. It would yield little or none at such partial pressures. In such annelids, however, the cells work at much lower partial pressures of oxygen—from 0 to 5 mm Hg. At these levels, their

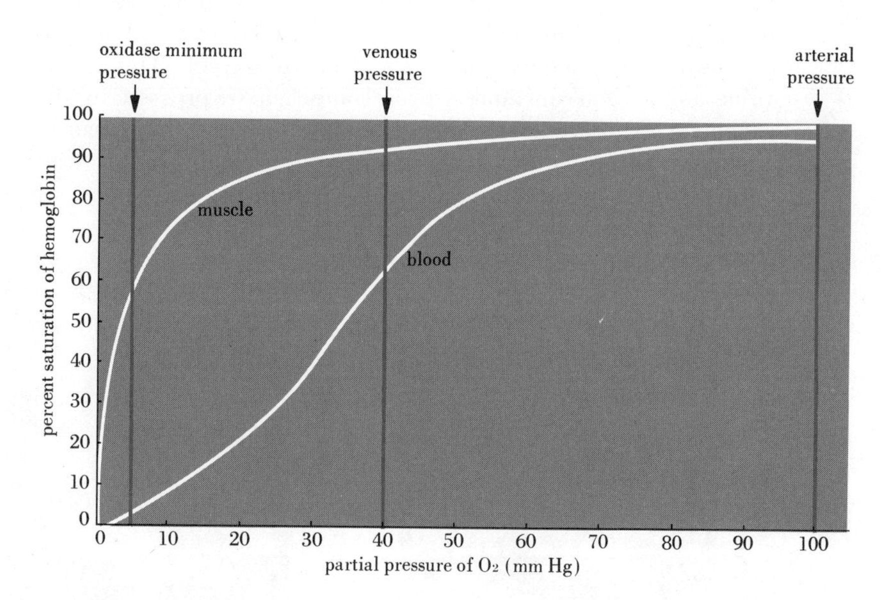


*Fig. 7-7 Oxygen equilibrium curves of mammalian hemoglobin and myoglobin measured at physiological temperature and pH. The vertical line at the left represents the level of oxygen at which the cellular cytochrome oxidase system appears to be saturated. Below this level the rate of tissue respiration falls off. The oxygen affinity of myoglobin (a molecule present in muscle in abundance) lies between that of hemoglobin and that of the oxidase system. Thus hemoglobin in the red blood cells yields oxygen to myoglobin in the muscle fibers nearby and myoglobin in turn yields oxygen to the enzyme systems using it. Thus we have delivery, storage, and utilization systems. (After Hoar, Wald, and Hill.)*

hemoglobin *will* yield oxygen. Such animals, of course, live in exceptionally oxygen-poor environments, in which the hemoglobin must have an affinity for oxygen at low partial pressures or it will pick up none at all.

Each curve representing the affinity of a hemoglobin for oxygen holds for only a given blood temperature, pH, and carbon dioxide content. Shifts in the latter two features, which characterize the lung (or gill) tissues versus the systemic capillary beds, tend to favor the uptake of oxygen at the respiratory surface and its surrender in the tissues. (See Fig. 7-8.)

Animals that are very active and that live where the partial pressure of oxygen is always high, such as birds and oceanic squids or fast-swimming fish, tend to have a kind of respiratory pigment (hemocyanin in squids) that gives off its oxygen at a fairly high partial pressure. Thus, the active cells can obtain oxygen rapidly along a steep concentration gradient from capillary to mitochondria (where the actual reactions with oxygen take place). Animals with respiratory pigments of this type must pay the price of vulnerability to low environmental oxygen partial pressures. Should they find themselves in a situation wherein the oxygen has been depleted by other organisms, their blood can take up but a small fraction of its full capacity of oxygen. This explains the vulnerability and unsuitability of some fish (expressed in short life expectancies) to the low oxygen concentrations that may occur in home aquaria. The best-suited fish for such living conditions are those that normally live under low partial pressures of oxygen, such as in hot, swampy waters.

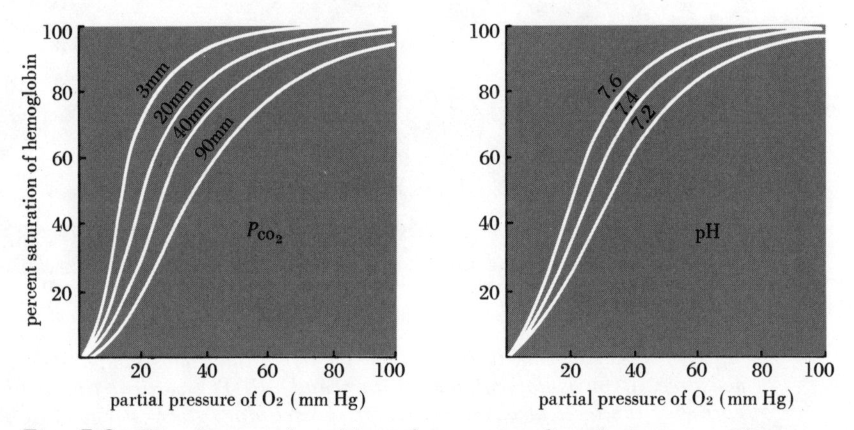


***Fig. 7-8*** *The shape and position of an oxygen dissociation curve (for man, for example) is affected by the pH and by the partial pressure of $CO_2$ in the blood. Both of these effects tend to increase the release of oxygen to active tissues which have high partial pressures of $CO_2$ and elevated pH. (After M. Gordon,* Animal Function: Principles and Adaptations, *Macmillan, 1968.)*

Hemoglobin and its role in oxygen transport are but two isolated examples of the functional adaptations that render almost every part of every animal an intricate mechanism of admirable efficiency. In an elementary account it is impossible to do justice to even a few of these biological phenomena, but the reader should bear in mind that every cubic micron of protoplasm, almost every molecule, is packed with such surprises.

## OTHER CONSTITUENTS OF BLOOD

The constituents of blood are not formed in the blood vessels; they must be transferred there from the sites of manufacture. The formation of blood cells, *hemopoiesis*, in adult mammals, takes place chiefly in the *bone marrow*. Almost all of the red blood cells, the platelets, and many of the leucocytes, especially types called *polymorphonuclear leucocytes* and the *monocytes*, are manufactured here. Under special conditions, the spleen and liver and—in some species and at some times of life, the kidneys—may engage in hemopoiesis. Lymphocytes are normally manufactured in specialized small organs, the lymph nodes, and in other lymphatic tissue, as noted elsewhere in this chapter. The number of erythrocytes and leucocytes in the blood is normally closely controlled and presumably monitored. In many species of mammals, the erythrocytes have a life of about 120 days. They may also be lost prematurely by hemorrhage, menstruation, or hemolysis by toxins. Nevertheless, the number is rapidly restored to the normal level. Evidence suggests that erythrocyte control is based on monitoring the oxygen-carrying capacity of the blood, since the number of erythrocytes increases, for example, at high altitudes (where the partial pressure of oxygen is low). It is not known how the number of leucocytes is monitored or regulated, but the population is remarkably constant in health, increasing only with infectious disease, parasitic infestation, or injury. Specialized cells in the spleen and liver detect defective or overage red cells and ingest them. Most of the hemoglobin from such discarded cells is discarded in the bile, but the iron contained in the hemoglobin is usually conserved.

The water and ion constituents of blood are monitored and regulated by the hypothalamus (part of the forebrain) and the kidneys. Circulating hormone level is a balance between production (by a large variety of endocrine organs) and destruction (often by the liver) or excretion (by the kidneys). Some elements of the monitoring of circulating hormone levels are discussed in Chapters 9 and 10. The blood levels of glucose, amino acids, and other energy-yielding or building-block molecules also appear to be monitored and to be influenced by hormones such as insulin, adrenocortical hormones, and adrenalin. The details are not yet completely known.

Heat is normally distributed by the bloodstream. It is generated in working muscles (which like all machines yield heat as a by-product of their work), by shivering (in which the muscles do no useful work but produce even more heat), in brown fat (a tissue found particularly in hibernating mammals) and, in varying degrees, in all other metabolizing tissues, and, in the gut, by bacterial metabolism. The blood carries heat to other, cooler sites (perhaps nearer the surface or more exposed). Cooled blood is also normally delivered from the skin to the deeper parts of the body. Heat production and loss are controlled by the hypothalamus, often by readjustments of circulatory distribution.

## FURTHER READING

Adolf, E., "The Heart's Pacemaker," *Scientific American*, 216 (3): 32–37, 1967.

Barrington, E. J. W., *Invertebrate Structure and Function*. London: Nelson, 1967.

Gordon, M., *Animal Function: Principles and Adaptations*. New York: Macmillan, 1968.

Hoar, W., *General and Comparative Physiology*. Englewood Cliffs, N. J.: Prentice-Hall, Inc., 1966.

Laverack, M. S., *The Physiology of Earthworms*. New York: Macmillan, 1963.

Levey, R. H., "The Thymus Hormone," *Scientific American*, 211 (1): 66–77, 1964.

Macfarlane, R. G., and A. H. T. Robb-Smith, *Functions of the Blood*. New York: Academic Press, 1961.

Manwell, C., "Comparative Physiology: Blood Pigments," *Annual Review of Physiology*, 22: 191–244, 1960.

Moore, J. A. (ed.), *Physiology of the Amphibia*. New York: Academic Press, 1964.

Perutz, M. F., "The Hemoglobin Molecule," *Scientific American*, 211 (5): 64–76, 1964.

Ruch, T. C., and H. D. Patton, *Physiology and Biophysics*, 19th ed. Philadelphia: Saunders, 1965.

Ruud, J. T., "The Ice Fish," *Scientific American*, 213 (5): 108–114, 1965.

Wood, J. E., "The Venous System," *Scientific American*, 218 (1): 86–96, 1968.

"Functional Morphology of the Heart of Vertebrates, a Symposium," *American Zoologist*, 8: 177–229, 1968.

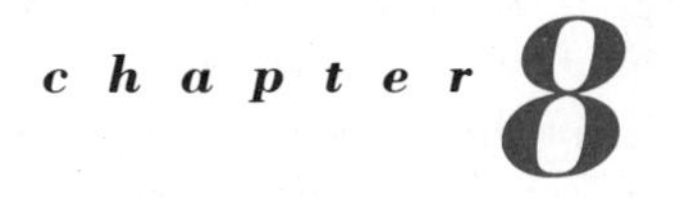

# Chemical Regulation

All highly organized animals are equipped with *excretory organs.* When these are numerous, small, and widely distributed through the body, they are usually called *nephridia*; in some animals they may be localized, as in the compact, but in many ways different, kidneys of mollusks, arthropods, and vertebrates. The term "excretory" is applied because all of these organs produce a fluid urine in which the animal disposes of some by-products of its metabolism, such as ammonia, $NH_3$ (usually as the ammonium ion), or urea $(NH_2)_2\ C{=}O$. But excretion of metabolic by-products is only one aspect of a much more fundamental regulatory function of these organs. *Urine* is a solution generally similar to blood, lymph, or tissue fluids that either bathe or flow close to the cells of excretory organs, but the composition of urine is adjusted to differ from that of the blood or tissue fluids in a manner that corresponds to the chemical needs of the animal. Substances that are too abundant may be more concentrated in urine than in blood; those that are in short supply may be more dilute in the urine than in the blood. Furthermore, these differences between blood and urine are not fixed, but vary from time to time according to changing chemical conditions inside the animal or in its surroundings. Thus, we may appropriately refer to these organ systems as functioning for regulatory excretion.

Living organisms first arose in the ocean and the internal fluids of all of the simpler marine animals now extant somewhat resemble sea water in the proportions of the major ions, $Na^+$, $K^+$, $Ca^{2+}$, $Cl^-$, and $SO_4^{2-}$. The concentrations of these ions in the blood of such primitive organisms may actually be even closer to the concentration of ions in the sea water of ancient times, during which their ancestors evolved. (See Fig. 8-1.) $K^+$ tends to be more abundant and $Mg^{2+}$ less so in their body fluids than in modern sea water. But many animals of all levels of

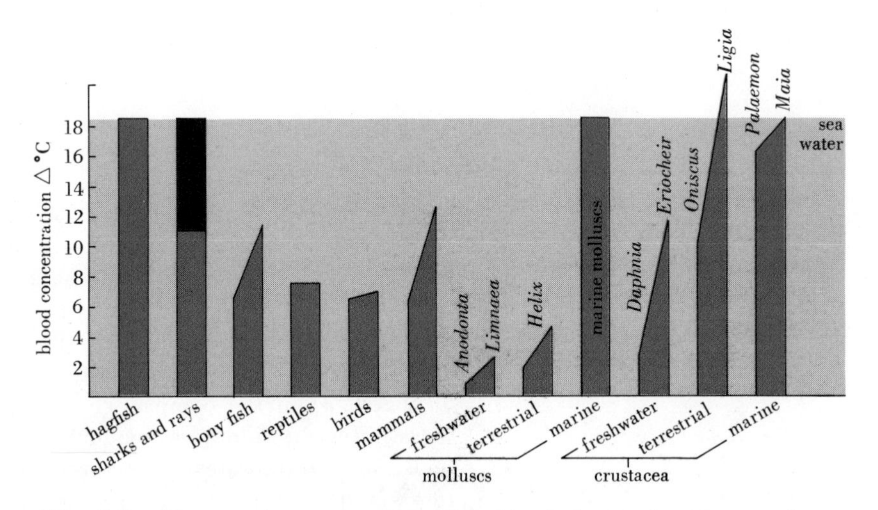


**Fig. 8-1** *The osmotic concentration of the blood of several* marine *vertebrates in contrast with molluscs and crustaceans from various habitats. Osmotic concentration is expressed as freezing point depression in degrees centigrade. Among the vertebrates, only the hagfish and the cartilaginous fish have blood isomotic with sea water. In sharks and rays, isosmosis is achieved largely by conserving urea and trimethylammonium (indicated in black) in the blood and tissues. Note as well that marine molluscs and crustacea may not have to expend much if any metabolic energy in regulating their blood osmotic concentration. They may, nevertheless, regulate the relative amounts of different ions. (Adapted from Lockwood, A. P.,* Animal Body Fluids and Their Regulation, *Harvard University Press, 1964.)*

complexity are able to live in freshwater streams and lakes and on land. Yet, their enzymes and other protoplasmic mechanisms still require a closely *regulated ionic environment.* This freedom to depart from the ancestral sea water requires that an animal regulate the composition of its fluids as a simple marine animal need not do. The matter was put succinctly by the nineteenth century physiologist Claude Bernard: "*La fixité du milieu intérieur est la condition de la vie libre*" (Constancy of the internal environment is essential for an independent life).

This sort of internal regulation of the animal body is an important example of what is now designated *homeostasis*, meaning the dynamic maintenance of a *constant* (or nearly constant) *state.* It is one of the unique attributes of living cells and organisms that they achieve homeostasis so well and with such compact mechanisms. Many essential types of homeostatic regulation are performed by individual cells that respond to variations in their environment by taking corrective action to offset departures from the most favorable state of the variable in question. The kidneys of animals perform simultaneously several independent types of homeostatic regulation by selective action on individual substances present in the blood.

Not all animals achieve the same degree of homeostasis and, indeed, for many of them homeostasis does not appear to be as central a theme as for others. The *osmotic pressure* of the body fluids, for example, of the marine annelid, *Arenicola*, almost matches that of the seawater medium in which *Arenicola* is experimentally immersed down to dilutions of about one part sea water to eight of distilled water. When animals respond to environmental change by drifting with it, they are called *conformers.* If the animal engages in metabolic work to oppose environmental changes, by maintaining its osmotic pressure and ionic proportions against concentration gradients, for example, it is called a *regulator.* Regulation and conformity apply as well to other environmental features such as temperature. Regulation need not be expressed only at the cellular level by increasing, for example, the uptake of $Na^+$ by active transport but may also be expressed by behavioral acts such as moving into the shade or into surrounds of higher humidity or higher moisture content. Where understood, regulation requires some device for measuring the item to be regulated and, in more complex animals, involves the nervous system and the endocrine glands. One can easily picture the metabolic problems facing migratory fish such as salmon or eels in regulating in both fresh and sea water. Many aquatic organisms, furthermore, inhabit estuaries or equivalent sites, where the salinity varies continually and between wide limits depending on tides, rainfall, the melting of snow, and other factors. The known details of regulation and conformity under such trying circumstances can be found in textbooks on comparative physiology.

## *OSMOTIC REGULATION IN FRESH WATER*

Consider the basic problem of any animal, be it protozoan or fish, that suddenly finds itself in fresh water, where the concentration of ions is much lower outside its body than in its tissue fluids. Parts of its surface are permeable to small molecules and ions (as a very minimum it must have a respiratory surface), and hence both ions and water itself tend to move along the concentration gradients that are necessarily present across these permeable surfaces. This process of moving across a permeable membrane is called osmosis. Salts will be lost and water will diffuse inward, but the latter process is usually much the more rapid of the two, because water penetrates most living membranes quite easily, and because, in addition to small ions, the internal fluids contain many larger molecules that cannot diffuse outward at all or only at negligible rates. Hence the inward diffusion of water and the outward diffusion of ions and small molecules both lead to progressively greater dilution of the internal fluids. (See Fig. 8-2.) If unchecked, this process produces swelling and eventually the death of the animal.

All freshwater animals effect osmotic homeostasis by carrying out some kind of regulatory excretion that "bails out" the excess of water as it diffuses inward. The excess water must be eliminated but ions must be conserved; therefore, urine must be excreted that is more dilute than

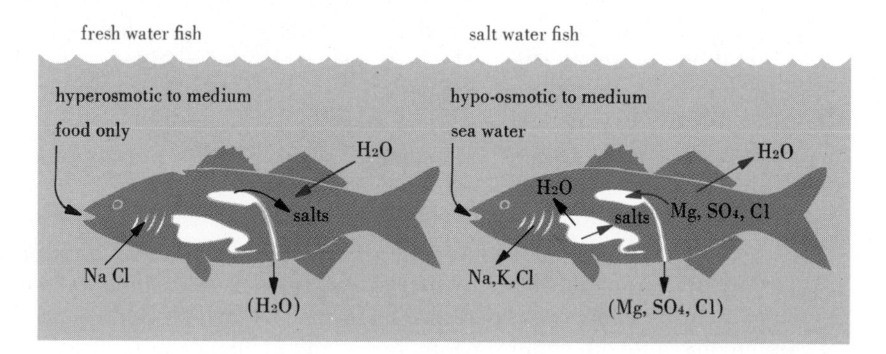


**Fig. 8-2** *A diagram of the main pathways of ion and water movement in freshwater and marine bony fish. Freshwater fish avoid swallowing water but nevertheless take up a considerable amount through their permeable surfaces by passive diffusion. This water is removed as a copious, dilute urine. Sodium ions, in particular, are actively secreted into the blood from the water by gill tissue. Salts are also conserved by active transport in the kidney. In marine fish, salt water is swallowed and absorbed to make up for water which is continually drawn from the body by the higher osmotic pressure outside. Monovalent ions are actively secreted into the water by gill tissues while divalent ions are excreted by the kidneys. Both groups have perpetual osmoregulatory problems. (From Prosser, C. L. and F. A. Brown, Jr.,* Comparative Animal Physiology, *2nd Ed., Saunders, 1961.)*

the blood or internal fluids. The separation of such a dilute solution from the internal fluids requires active chemical work on the part of specialized cells usually gathered together as an excretory organ. So important is this regulatory excretion that under some circumstances it accounts for the consumption of an appreciable fraction of the total metabolic energy of the animal.

## *THE STRUCTURE OF EXCRETORY ORGANS*

Even Protozoa have an organelle, called the *contractile vacuole*, for regulatory excretion. This is a fluid-filled vesicle, in the cytoplasm, that expands slowly from time to time and then empties to the outside. Such vacuoles contain a more dilute solution than the cytoplasm. Contractile vacuoles are formed and emptied much more frequently in freshwater protozoans than in their close relatives that live in sea water. Some Protozoa can live in a wide range of salinities; the more dilute the solution in which they are studied, the more active are their contractile vacuoles. The exact mechanism for filling the contractile vacuole is not known, but its wall is a membrane only 60 Å thick. The cytoplasm close to this membrane contains a layer of mitochondria and a dense layer of vacuoles 200–2000 Å in diameter that probably open into the vacuole to fill it. Since mitochondria are known to contain many of the enzymes concerned with the stepwise breakdown of food molecules to yield energy, it is reasonable to assume that they are providing the energy for the chemical work of separating from the cytoplasm the dilute "urine" to fill the contractile vacuole.

Although the contractile vacuole is an important intracellular organelle for regulatory excretion of water, regulation in many cells occurs in the absence of any recognized structural specialization. A number of fresh-water protozoans do not have contractile vacuoles; nor are such vacuoles found in fresh-water coelenterates, which nevertheless maintain a higher salt concentration than that of the surrounding water. All of the flatworms, most of the roundworms, and all of the annelid worms have some sort of specialized excretory organs that consist of tubules emptying through pores on the surface of the body. These *excretory tubules* often branch; their inner ends are thin-walled and located close to blood or to tissue fluids such as those in the body cavity. In the flatworms, the inner end of each tubule consists of a single specialized cell with a tuft of cilia that beat continuously and set up a gentle current along the tubule toward the outside. (See Fig. 8-3). These cells are called *flame cells* because, under the microscope, the actively beating cilia appear to flicker like a candle as they move. In annelid worms, the excretory tubules begin with a funnel-shaped

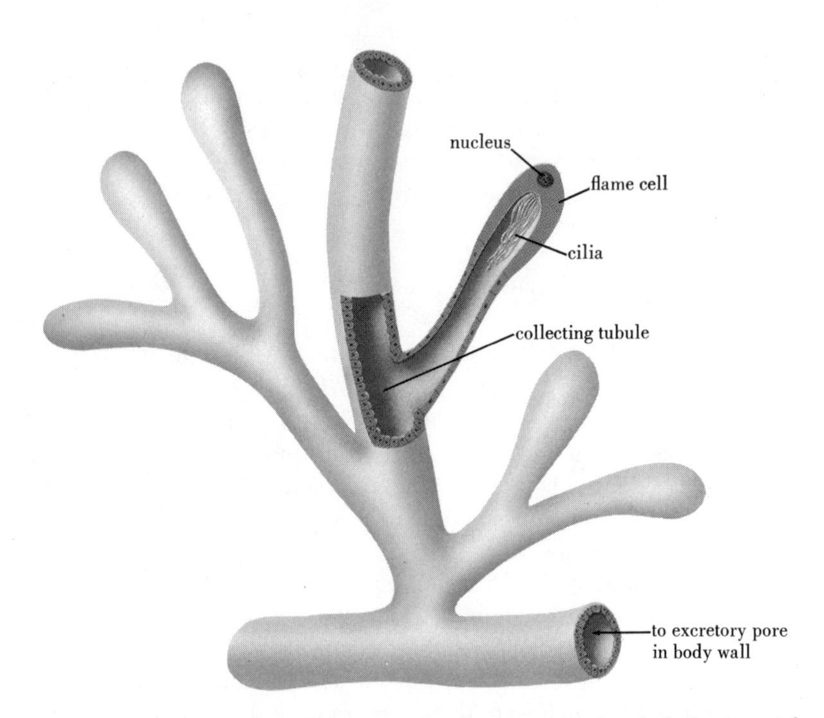


***Fig. 8-3*** *A portion of the excretory of a flatworm. Each tubule begins with a flame cell. The tuft of cilia, the beating of which has been likened to a candle flame, sets up a current in the fluid within each tubule.*

opening, from one segment of the body cavity, which leads to a long convoluted tubule, which, in turn, drains, via an external pore, to the outside of the animal's body.

In mollusks and arthropods there are similar excretory tubules, but they are usually longer and are concentrated into definite kidneys that empty to the outside through larger collecting tubules. In some crustaceans, urine is collected in a thin-walled *bladder* before being discharged to the outside through two pores located on the head. Although the inner ends of the tubules usually do not make a direct connection to the body cavity, they are closely surrounded by the blood of the open circulatory system. Insects differ from crustaceans in having as excretory organs a series of long, thin *Malpighian tubules* ramifying through the body and emptying into the posterior end of the intestine. The two types of excretory systems found in arthropods are shown in Fig. 8-4.

Vertebrate animals have paired *kidneys*, each consisting of thousands of *kidney tubules* or *nephrons*, which are several millimeters long or more and are, in higher vertebrates, fitted together in a highly organized pattern. (See Fig. 8-5.) All begin with a tubule that is in close contact with a network of capillaries, and all converge, finally, into a

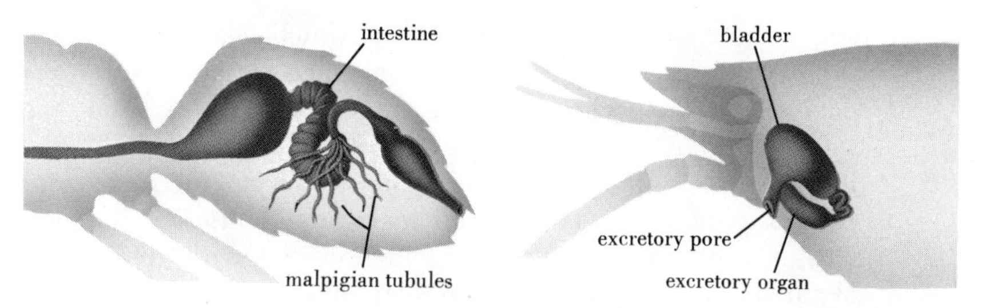


***Fig. 8-4*** *The excretory system of an insect* (left) *and a crustacean* (right). *In freshwater crustaceans, a dilute, fluid urine is excreted directly into the water. In insects, the Malpighian tubules (which make up the excretory system) empty into the digestive tract. In many cases, uric acid is the chief nitrogenous compound of the "urine." Uric acid, which is only slightly soluble in water, precipitates out as crystals and is excreted in solid form. Water is conserved (important to many terrestrial animals) but a peristaltic mechanism is usually required to move the sludge along.*

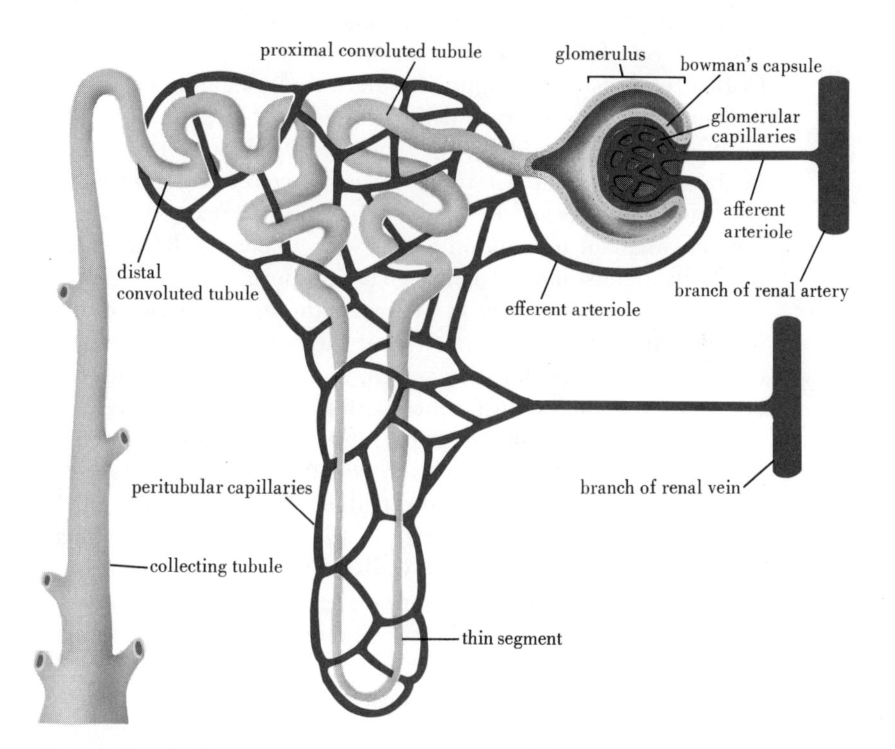


***Fig. 8-5*** *A diagrammatic representation of the nephron of a mammal. Filtration occurs through the walls of the glomerular capillaries and Bowman's capsule. The filtrate provides the raw material for urine production. As the fluid moves along through the lumen of the nephron, the water and solute concentrations are continually being adjusted by a variety of passive and active processes. The peritubular capillaries probably bear a much more regular relationship to the tubule than has been shown here but reconstruction of the exact geometry is, as yet, incomplete.*

single duct on each side, the *ureter,* which leads, in mammals, to a *urinary bladder.* This in turn is connected to the outside by a single duct, the *urethra.* The vertebrate nephron and the excretory tubules of some other phyla are in intimate association with the circulatory system. In vertebrates, this association takes two forms, the *glomerulus* at the very beginning of the nephron and the *peritubular* capillaries that cluster around the tubule along much of its length until it joins other nephrons to empty into the *collecting tubules.*

Each glomerulus is a tuft of capillaries supplied by an afferent arteriole that arises as one of many branches of the renal artery (the blood supply of the kidney). The walls of these glomerular capillaries are very thin. The electron microscope reveals that they are penetrated by many pores too small to be resolved by the light microscope. Unlike most capillaries, those of a glomerulus, in mammals, coalesce to form a small efferent arteriole, with rather muscular walls, instead of a small vein. This efferent arteriole leads to a second set of capillaries that pass close to, and follow the path of, one or more nephrons. (See Fig. 8-5.) Only after passing through these peritubular capillaries does the blood flow into veins for a return trip to the heart. The glomerulus is enclosed by a capsule, formed by the pushed-in bulbous enlargement at the beginning of the nephric tubule. This *glomerular* or *Bowman's capsule* resembles a squeezed tennis ball that has collapsed so that both sides now lie adjacent to each other but with a space in between and with a generally hemispherical shape. The glomerulus is an arrangement by which fluids can leave the bloodstream through the walls of the glomerular capillaries, penetrate the wall of the nephron which is also the wall of the capsule and find themselves, then, in the beginning of the *lumen* or channel of the nephric tubule. Along much of the length of the kidney tubule there is further opportunity for exchange among the blood flowing through the peritubular capillaries, the extracellular fluid surrounding the tubule, the intracellular fluid of the cells of the tubule wall, and the fluid in the lumen of the tubule. Here exchange occurs across not only the capillary walls but across the cells forming the tubule as well. Recent discoveries suggest that deep grooves between the cells of the tubule may be very important sites of regulatory activity. These cells take an active part in such exchange.

Glomeruli are especially large and well developed in freshwater bony fish, whereas in marine bony fish they are smaller and less numerous. A few species of marine fish lack glomeruli altogether and have nephrons that are served only by peritubular capillaries. A question of marked interest is why there should be this correlation between the kind of water in which the fish live and the microscopic anatomy of their kidneys.

## THE FORMATION OF URINE

In the formation of urine in all excretory tubules that have been carefully studied, from those of flatworms beginning with a flame cell to those of mammals arising as a glomerular capsule, the fluid in the beginning or proximal end of the tubule is a solution very much like blood plasma or other body fluids. All these tubules are leaks in the animal's circulatory system or holes through which its body fluids tend to escape. The solution inside the tubule, however, is "processed" in remarkable ways as it moves along. Some substances are added and others removed. The end result is the excretion of a solution differing from the internal fluids in a manner that contributes to achieving a homeostasis of the internal environment. As far as is known, the same basic processes are at work in all types of excretory tubules; but these are more efficient, and much better understood, in the case of vertebrate animals. Consequently this elementary account of the mechanisms of urine formation will be confined to the vertebrate nephron.

The first step obviously occurs at the glomerulus. Analysis of fluid withdrawn from the lumen of the nephron near the glomerular capsule by micropipette shows that the only important difference between the fluid that here enters the lumen of the renal tubule and the blood plasma is the absence of blood cells and protein molecules. In fact, the constitution of the fluid is just what one would expect on the assumption that all but very large molecules pass, by filtration, through the pores, visible under the electron microscope, in the walls of glomerular capillaries. Furthermore, measurements of the rate at which this filtrate of plasma is exuded into the whole set of nephrons in a human kidney reveal the startling fact that more than 100 ml are lost from the circulatory system every minute. About 20 percent of the volume of all the blood flowing through the glomeruli is filtered into the renal tubules. The driving power for this process of filtration is the hydrostatic pressure in the glomerular capillaries deriving from the contraction of the heart muscle.

Obviously some compensating process must intervene, since the normal output of urine is only about 1 ml per minute. This compensating process is the reabsorption of water and the selective reabsorption of other molecules from the tubules back into the peritubular capillaries. Over 99 percent of the water in the *glomerular filtrate* is reabsorbed. Glucose and amino acids are almost totally withdrawn and returned to the blood and cannot normally be detected in the urine at all, although microanalysis of fluid in the tubule close to the glomerulus discloses their presence there. The rate of reabsorption of glucose is limited. If the plasma glucose level (and glomerular filtrate glucose level) are abnormally high—as often occurs in diabetes mellitus (a disease involving malfunction of the endocrine portion of the pancreas) or, less frequently, in normal young adults who have just consumed several sucrose-rich candy bars—the capacity of the glucose-transferring mechanism is exceeded and some glucose is left behind to be excreted in the urine. Glucose is, therefore, said to have a *transfer maximum.* Many of the small circulating molecules, especially nutrients, have similar transfer maxima. In some cases they may compete with each other for reabsorption; in other cases they do not, and therefore it is implied that there is more than one transfer mechanism and that each may be more or less specific. For example, several transfer mechanisms are apparently involved in the active reabsorption of amino acids from the lumen of the nephron. Arginine, lysine, and histidine share the same mechanism, since elevation of the plasma concentration (and, therefore, of the concentration in the lumen) of one depresses the reabsorption of the others. A second mechanism is responsible for the reabsorption of leucine and isoleucine. All of these mechanisms, as far as we now know, are contained in the cells of the tubule and involve the expenditure of energy, since the molecules are being moved against a concentration gradient. We have here further examples of *active transport* (for the conservation of valuable molecules). In fact, the existence of a transfer maximum or a limited capacity subject to saturation may be taken as evidence for the presence of active transport.

On the other hand, only about half of the urea present in the glomerular filtrate is reabsorbed. Thus, the net concentration of urea in the urine is greater (when 99 percent of the water has been reabsorbed) than that in the blood plasma. There is no evidence for active reabsorptive transport of urea. Some urea diffuses back into the peritubular capillaries along with water that is reabsorbed, following the concentration gradient. This is called *passive reabsorption.* In advanced renal disease, the excretion of urea is impaired by markedly reduced glomerular filtration. Tubular function is also disturbed.

Some substances are also moved directly from the peritubular capillary blood into the lumen of the nephron. In amphibians, the peritubular capillaries derive from the renal portal vein, whereas the glomerular capillaries are supplied by the renal artery. One can, therefore, experimentally tie off the renal arteries, thereby eliminating glomerular function. There is, then, no glomerular filtrate. If such a substance as phenol red is then injected into the renal portal vein, it will be found in high concentration in the renal tubules. It is *actively secreted* into the lumen of the tubule by the cells of the tubule wall. In man, penicillin is also such a substance. Penicillin passes into the glomerular filtrate freely. The 80 percent that remains in the peritubular blood, however, is actively secreted into the lumen of the tubule. All of the penicillin arriving in the renal arterial blood is removed either by filtration or by secretion and is totally excreted in the urine. It is said to be completely *cleared* from the blood plasma. Alternately one may say that the *blood plasma clearance* for penicillin is 100 percent. (See Fig. 8-6.)

Many drugs are actively secreted by the tubules. The administration and dosage of these drugs must be carefully assessed when tubular function has been impaired by renal disease. Some substances —such as the polysaccharide, insulin—are filtered but are neither passively nor actively secreted or reabsorbed. By choosing molecules such as those mentioned (which are filtered or not and which are actively reabsorbed or secreted) and by using them experimentally, monitoring their appearance in the urine and their changing concentration, one can unravel much of renal physiology.

Water absorption is passive. When the osmotic pressure of the fluid in the tubule is reduced by the active reabsorption of glucose,

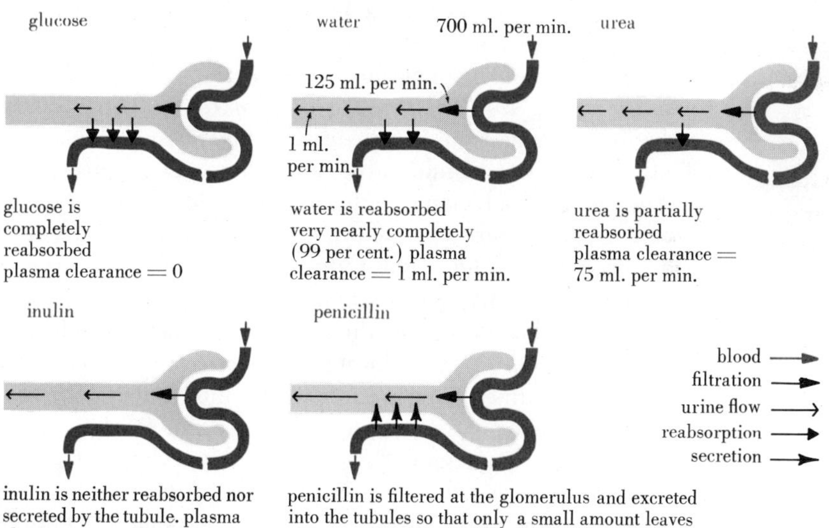


**Fig. 8-6** *A diagrammatic representation of* clearance. *In this example, renal plasma flow is 700 ml. per minute (the red and white cell volume are excluded). Glomerular filtration amounts to about 20 percent of renal plasma flow or about 125 ml. per minute. The filtrate includes all the solutes contained in plasma except for proteins. Almost all the water is reabsorbed. Thus, if none of a particular solute is reabsorbed, about 125 ml. per minute of plasma will have been* cleared *of this substance. Such is the case for inulin. The other examples have been chosen to illustrate other levels of clearance. (Adapted from Bell, G.H., J.N. Davidson, and H. Scarborough,* Textbook of Physiology and Biochemistry, *54th Ed., Williams & Wilkins, 1961.)*

amino acids, and $Na^+$, water diffuses passively into the now more concentrated interstitial fluid (and thence into the blood plasma). Ions such as $Na^+$ and $K^+$ are usually conserved. At least, their concentrations are adjusted homeostatically. The active reabsorption of $Na^+$ in the *proximal tubule* is accompanied by passive reabsorption of water. Such a mechanism reduces the urine volume but cannot substantially affect the osmotic pressure. Nevertheless, the kidneys of mammals and birds do elaborate a hypertonic urine, which is more concentrated than blood plasma. (The tonicity of a solution is related to its osmolarity or osmotic pressure. An isotonic solution is one that shows no net exchange of water with intact erythrocytes. Erythrocytes swell in a hypotonic solution and shrink in a hypertonic solution. In many but not all cases, an isotonic solution is also isosmotic, a hypotonic one is hyposmotic and a hypertonic one is hyperosmotic.)

The progressive changes in water and ion concentration in the tubule are systematized, sequential, and geometrically arranged so as to set up a *countercurrent multiplier*, which ultimately conserves water and permits the excretion of a hypertonic urine. (See Fig. 8-7.) The geometric arrangement of the parts of the nephron is not random or happenstance. The proximal and distal portions of the tubule are connected by a long *thin segment* (or *loop of Henle*). The thin segment is in the form of a hairpin, its proximal and distal arms lying adjacent to each other but with the fluid in the lumen flowing in opposite directions. The thin segments of many nephrons are lined up side by side. Their path and orientation are also followed by the collecting ducts of the same part of the kidney. The countercurrent multiplier system depends on this double-hairpin shape of the nephrons.

The physiological key to the countercurrent multiplier system is that the distal loop of the thin segment *actively transports* $Na^+$ from the fluid of the lumen into the interstitial fluid but is *impermeable to water*. Thus, the osmotic pressure of the urine in the lumen of the distal arm of the thin segment is progressively reduced and that of the interstitial fluid is raised. The proximal arm of the thin segment is immediately adjacent to the distal arm and is permeable to both $Na^+$ and water. Sodium ions diffuse passively, therefore, into the lumen of the proximal arm and water diffuses with them. The fluid that originally flows into the proximal arm of the thin segment from the previous portion of the nephron is isotonic. As it flows through the proximal arm of the thin segment, it picks up sodium ions and becomes progressively more hypertonic. As it flows through the distal arm, sodium ions are withdrawn and the fluid becomes isotonic or even hypotonic. The proximity and counterflow of the fluid columns in the two arms of the loop plus the active transport of $Na^+$ and the water impermeability of the distal arm set up a countercurrent multiplier system. In addition (and essential to the function), a gradient of tonicity becomes established in the interstitial fluid, which is now hypertonic at the apices of the loops and progressively more isotonic toward the open end of the loop. Finally, the urine passes through this gradient again when it reaches the collecting duct. (See Fig. 8-7.) The membrane of the collecting duct may be

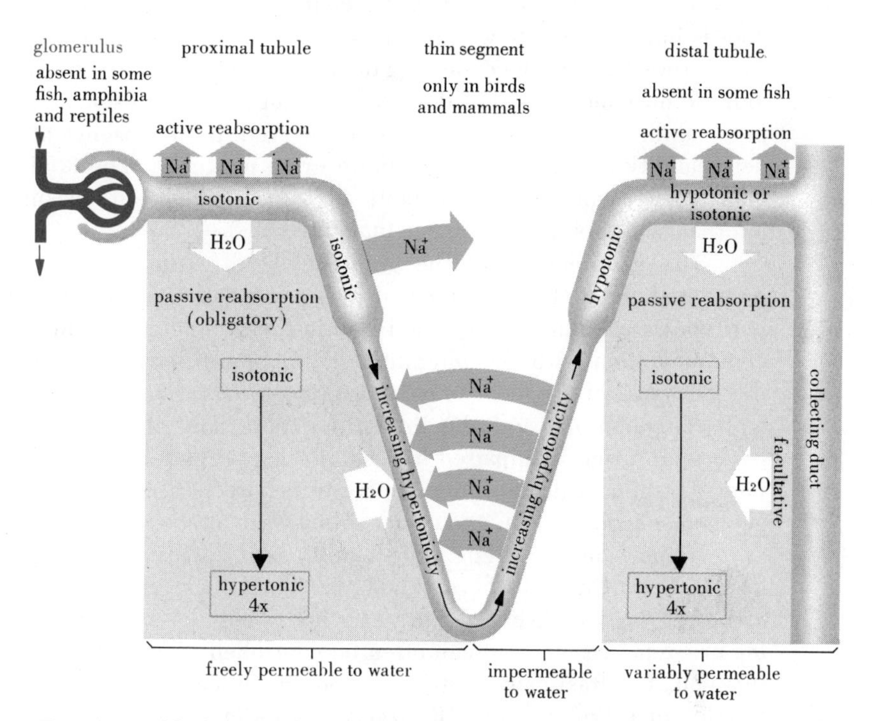


***Fig. 8-7*** *Mechanism for producing either concentrated or dilute urine in mammals. $Na^+$ is actively absorbed in the proximal tubule, carrying water with it passively. Thus, the fluid reaching the thin segment is reduced in volume but isotonic with plasma. The two arms of the thin segment lie side by side. Let us skip over the proximal arm for a moment. In the distal arm, sodium ions are again actively absorbed but the tubule is impermeable to water. Thus the fluid in the lumen becomes increasingly hypotonic as $Na^+$ is withdrawn. The tissues surrounding the thin segment (or, in reality, thousands of similarly arranged thin segments) also show this gradient of tonicity. Indeed, sodium ions also diffuse passively back into the proximal arm of the thin segment lying nearby and as they diffuse in they carry water in with them passively. The two arms form a counter current system such that the sodium ions become most concentrated at the bend and most dilute at the "free" ends. The key features required for this system to work are impermeability to water of the distal arm of the thin segment, active transport of $Na^+$, and precise geometric relationships. The "purpose" of the counter current design appears to be to establish the gradient of tonicity surrounding the collecting tubules. The wall of the collecting tubule may be either permeable or impermeable to water ($Na^+$ is not acted upon here). If permeable, water is passively reabsorbed, because of the hypertonicity of the surrounding tissues, and a concentrated urine is excreted. Water is conserved. If the wall is impermeable, no water is reabsorbed; a dilute urine is excreted. The permeability of the collecting tubule to water is altered by antidiuretic hormone secreted by the hypothalamus in response to osmotic monitors. (After W. Hoar,* General and Comparative Physiology, *Prentice-Hall, 1966.)*

permeable to water and to $Na^+$. If the collecting duct is permeable to water, water will be withdrawn passively by the progressively more concentrated surrounding interstitial fluid. The urine produced will be as hypertonic as the most concentrated interstitial fluid. If the collecting duct is impermeable to water, a hypotonic or isotonic urine is excreted. The permeability of the collecting duct to water is regulated by the antidiuretic hormone (ADH) of the hypothalamus. In the presence of ADH, permeability is increased and water conserved. In the absence of ADH, permeability is inhibited and a dilute urine produced. The production of ADH is responsive to the osmotic pressure of the blood measured by internal monitors (probably hypothalamic cells).

The maintenance of the gradient of $Na^+$ concentration in the interstitial fluid of the kidney also depends on the integrity of the peritubular capillaries since the blood in these capillaries is in osmotic communication with the interstitial fluid. Some portions of the peritubular capillaries appear to be arranged as countercurrent exchangers with the nephrons and the interstitial fluid. The ability to produce hypertonic urine is impaired in renal disease. Probably all of the components—glomerular filtration, active transport, peritubular blood flow, geometric arrangement, and permeability—are affected to some extent.

If the countercurrent system fails, as it apparently does in congestive heart failure, where venous pressure is elevated, sodium ions are excessively reabsorbed and water is carried along passively. Swelling or edema results and can often be controlled by correcting cardiac function, by drastically restricting sodium intake, or by using drugs which cause the retention of large amounts of water in the urine (by largely unknown mechanisms) and, in effect, flush out the edema, carrying various and sundry desirable ions and molecules out, too, in the rush.

It is surprising at first thought that our kidneys and those of most other vertebrates should operate in the manner just described. Why should such a large potential leak from the blood system be tolerated? The answer is believed to lie, in large part, in our evolutionary history as the descendants of ancient fishes. Bear in mind the basic osmotic problem of freshwater animals—removal of the water that diffuses in from outside. We can readily imagine that kidneys equipped with glomeruli would be effective in removing such excess water. Having separated a considerable volume of blood plasma by filtration across the very thin capillary walls of the glomerulus, the rest of the nephron has the function of selectively reabsorbing much of the water and various dissolved molecules and ions. The total salt concentration of the blood and tissue fluids of freshwater fishes is somewhat less than half that of sea water, though of course it is still well above the concentration of salt in fresh water. Apparently active reabsorption of solutes is easily designed but the active reabsorption of water is not. Therefore the tubule is designed primarily around the water problem.

These considerations also help to explain why marine fish should have small glomeruli or none at all. Their body fluids have nearly the same total salt concentration as freshwater fish and, indeed, all vertebrates are similar in this respect. Thus, being in an environment that is hyperosmotic relative to their blood, they tend to lose rather than gain water. The kidneys are not needed to "bail out" water but to regulate ionic balance and to get rid of nitrogenous waste products. The muscles in the walls of the arterioles supplying the glomeruli and in the venules (in fish) draining the glomeruli also control the amount of blood flowing through them. This may be especially important in fishes that move back and forth between fresh and salt water. In other words, the vertebrate glomerular nephron seems designed for the needs of freshwater fish; this observation, together with other evidence, leads biologists to believe that the earliest bony fishes lived in fresh water and that present-day marine fish, as well as terrestrial vertebrates, are descended from them.

The cartilaginous fish have long been known to have a curious and different solution to their problem of water retention in a hyperosmotic environment. Instead of relying directly on water conservation, the cartilaginous fish retain concentrations of urea in their body fluids sufficient to raise the osmotic pressure close to that of the sea water in which they live. Thus, they continue to have the problem of ionic regulation but do not have the problem, directly, of osmotic regulation. The recently discovered lobe-finned fish, *Latimeria*, an offshoot of a line which is believed to have given rise to tetrapods, has been shown also to function with a high concentration of tissue urea, perhaps for the same purpose.

Quite apart from these historical reasons for our kidneys being what they are, the system of filtration and selective tubular reabsorption and secretion has certain advantages from the point of view of regulation, basically the most important aspect of kidney function. Let us consider the regulation by the kidney of the blood levels of a few typical substances, beginning with water itself. Since over 99 percent of the water in the glomerular filtrate is ordinarily reabsorbed, a very small change in this reabsorption will produce a large change in the total rate of water loss; for example, a shift from 99 to 98 percent reabsorption means a doubling of the water output. Thus the needs of the animal for water elimination or water conservation can be met by slight alterations in the rate of reabsorption. An extreme example of the need for water conservation would be certain desert rodents whose only regular water intake is that derived from the metabolic breakdown of carbohydrates, proteins, and fats in the dry seeds on which they feed. They produce a very hyperosmotic urine. Another important example is the regulation of the potassium ion, the concentration of which must be kept within certain limits in the blood, since it is required for the normal functioning of all cells and especially of nerve and muscle (including cardiac muscle) cells. Because foods vary widely in their potassium content, the blood level tends to fluctuate with diet. In the nephron, most of the potassium ions are reabsorbed along with the water, but the exact quantity varies with the concentration of potassium ions in the blood; if this is higher than normal, fewer ions are reabsorbed and more are lost in the urine. Potassium may also be actively secreted. The mechanism for such adjustments in the rate of ion reabsorption is not known at present. Similarly, other ions are reabsorbed at different rates to achieve a homeostatic regulation of their level in the blood stream. High rates of glomerular filtration may also function for the "easy" elimination of toxins. Even if these are neither reabsorbed nor actively secreted they will, nevertheless, be rapidly cleared from the plasma.

Not all regulation of osmotic pressure and relative ion concentrations in vertebrates is handled by the kidneys. Marine bony fish, which continually lose water to the more concentrated sea water, compensate for this loss by drinking sea water and secreting salts extrarenally. They are, thus, left with a bonus of fresh water. Monovalent ions, in particular, along with some nitrogenous wastes, are secreted by cells of the gills. Marine bony fish kidneys apparently function principally to excrete divalent ions such as $Mg^{2+}$ and $SO_4^{2-}$. (See Fig. 8-2.)

Among tetrapods, a fairly wide variety of birds and reptiles living in marine environments or living where deprived of fresh drinking water, depend upon salt-secreting *nasal glands*. Thus marine birds such as albatrosses or terns can drink sea water and secrete the NaCl in high concentrations within a few minutes. By metabolic work they are, in effect, distilling sea water and retaining the distillate. The saline secretions may be seen dripping from the beak of many sea birds shortly after they drink. Other accessory systems for regulating salt and water balance are known in other animals.

## *NITROGEN EXCRETION*

Most animals take in with their food more amino acids and other compounds containing nitrogen than they require either as food or in the synthesis of new proteins for repair or growth. As a result, excess nitrogen usually must be eliminated by excreting a variety of nitrogenous compounds. A few animals may store excess nitrogen in a harmless or inactive form such as guanine. A very common by-product of amino acid metabolism in the cells of all animals is ammonia, which in aqueous solution takes the form of the ammonium ion, $NH_4^+$. This is readily soluble and, because it is rather toxic, its concentration cannot be allowed to rise. $NH_4^+$ is the chief vehicle for the excretion of excess nitrogen in most invertebrate animals except the insects and some of the terrestrial snails. It is also used by fishes and the aquatic stages of amphibians, crocodiles, and aquatic turtles, all of which have an abundant supply of water and excrete a copious, dilute urine.

Terrestrial vertebrates and marine fishes have problems of water conservation, and it is apparently difficult for them to spare enough water to dilute the ammonium ions arising from excess nitrogen intake. Instead, their nitrogen excretion is achieved in the form of urea or other nitrogenous substances that are much less toxic than $NH_4^+$. In addition to urea, many marine fish excrete considerable quantities of trimethylamine oxide, which gives rise to one of the characteristic odors of dead fish. The most interesting adaptation of nitrogen excretion to terrestrial life and its attendant problems of water conservation is the conversion of nitrogenous products into the nearly insoluble substance, uric acid, a large molecule containing four NH groups. Uric acid crystals make up the semisolid, whitish paste excreted by birds, snakes, and lizards. Almost no water is lost. In the case of birds, there is a substantial savings in weight, since the water that would otherwise have to be carried to convey urea in the urine need not be stored. The low solubility of uric acid means that it does not substantially raise the osmotic pressure of the urine and, thus, alter another of the problems of urine production. Since uric acid is, however, excreted as crystals, special provisions must be made for keeping the solid material moving along the urinary tract. Mammals do not normally excrete uric acid. The advantage they gain by secreting urea rather than uric acid may be in more completely utilizing the energy content of the nitrogen-containing molecules they ingest. Terrestrial snails excrete mostly uric acid, and many insects excrete it, as well, via their Malpighian tubules; in other insects, uric acid crystals are deposited in various parts of the body—for example, as the white scales on the outer surfaces of the wings of certain white butterflies.

These adaptations to fit the nitrogenous excretory products to the animal's way of life are achieved to a great extent by cells of the liver and kidney, which convert excess amino acids and other nitrogen-containing compounds into various proportions of $NH_4^+$, urea, trimethylamine oxide, or uric acid, any of which may be excreted by their kidneys. Not only do all these cells achieve an immediate type of homeostasis, but they may change the entire direction of their biochemical work according to the needs of the animal. This is most clearly exemplified by the embryos of snakes, lizards, and birds, which must develop from one cell into a complete animal without any water except that present in the original egg. During the first few days of their development, $NH_4^+$ is produced. Some of it escapes through the shell as gaseous ammonia. Later, as their organ systems become more fully formed and effectively functional, urea and uric acid predominate in turn. Since no liquids or solids can be extruded to the outside, uric acid crystals simply accumulate just inside the shell.

There are many other known cases in which kidneys or other organs adjust their activities in such a way as to achieve homeostasis. Still other phenomena of this type await adequate investigation. The foregoing examples suffice to demonstrate the complex web of interactions that are required to maintain the harmoniously organized internal environment of a living animal.

## FURTHER READING


Chew, R. M., "Water Metabolism in Desert-inhabiting Vertebrates," *Biological Reviews of the Cambridge Philosophical Society*, 36: 1–31, 1961.

Craig, R., "The Physiology of Excretion in the Insect." *Annual Review of Entomology*, 5: 53–68, 1960.

Edney, E. B., *The Water Relations of Terrestrial Arthropods.* New York: Cambridge, 1957.

Florey, E., *An Introduction to General and Comparative Physiology.* Philadelphia: Saunders, 1966.

Hoar, W., *General and Comparative Physiology.* Englewood Cliffs, N.J.: Prentice-Hall, 1966.

Laverack, M. S., *The Physiology of Earthworms.* New York: Macmillan, 1963.

Lockwood, A. P. M., "The Osmoregulation of Crustacea," *Biological Reviews of the Cambridge Philosophical Society*, 37: 257–305, 1962.

———, *Animal Body Fluids and Their Regulation.* Cambridge, Mass: Harvard University Press, 1964.

Martin, A. W., "Comparative Physiology (Excretion)," *Annual Review of Physiology*, 20: 225–242, 1958.

Nicol, J. A. C., *The Biology of Marine Animals.* New York: Interscience, 1960.

Pitts, R. F., *Physiology of the Kidney and Body Fluids.* Chicago: Year Book Medical Publishers, Inc. 1963.

Potts, W. T. W., and G. Barry, *Osmotic and Ionic Regulation in Animals.* New York: Pergamon, 1964.

Schmidt-Nielsen, K., "The Salt-Secreting Gland of Marine Birds," *Circulation*, 21: 955–967, 1960.

Shaw, J., and R. H. Stobbart, "Osmotic and Ionic Regulation in Insects," *Advances in Insect Physiology*, 1:315–399, 1963.

Smith, H. W., *Principles of Renal Physiology.* New York: Oxford, 1956.

Solomon, A. K., "Pumps in the Living Cell," *Scientific American*, 207 (2): 100–108, 1962.

# chapter 9

# Reproductive Systems

Animals living today are here because their ancestors succeeded in reproducing themselves more effectively than did other competing forms of life. By far the most common way in which new individuals of the same species are produced is *sexual reproduction*; two cells, called *gametes*, fuse to form a one-celled *zygote*, which in turn divides repeatedly to grow into a fully developed, new individual. The basic phenomena that underlie this vital process of reproduction are treated in *Genetics*, *Development*, and *Evolution* in this series, as well as in *Cell Structure and Function* by Loewy and Siekevitz; hence this chapter will be confined to discussing reproductive behavior and to dealing with the organ systems of animals that produce and regulate the production of gametes, facilitate their coming together, and serve to protect and nourish the resulting progeny.

The organs in which gametes are produced are called *gonads*; in almost all metazoan animals there are two kinds of gametes. A large type of gamete called an *ovum* (plural *ova*) contains stored food materials (mostly fat, some proteins, minerals, and a complete collection of vitamins) that will nourish the developing zygote. The smaller and far more numerous type, called *spermatozoa* or *sperm*, swim actively by means of a flagellum. Spermatozoa contain very little stored food. The gametes are described in detail in *Development*. Usually the two types are produced in different types of gonads; most commonly an adult animal is either a male (with *testes*, which produce sperm), or a female (with *ovaries*, which produce ova). There are exceptions to this rule. In many invertebrates, especially in the phylum Platyhelminthes, ovaries and testes occur in the same animal, which is then called a *hermaphrodite* (see Figs. 2-2 and 2-3). True hermaphrodites, with simultaneously functioning ovaries and testes, are also known among vertebrates. Several genera of marine bony fish, for example, are normally hermaphroditic (but not self-fertilizing), alternating roles in courtship and mating; that is, one will assume typically female behavior and shed eggs while the other will fill the masculine role behaviorally and shed sperm. They may then, immediately afterwards, reverse roles. How agreement is reached on which will fulfill which role in a given courtship sequence is not fully established. The fish with the fuller contours, however, representing greater egg content, usually takes the female role. Hermaphroditism in reptiles, birds, or mammals has never been confirmed. In some species of birds, however, and in some fish, as well, sex reversal, with conversion of the ovary into a testis, may occur with advancing age.

There are also several types of *asexual reproduction*. In some coelenterates and flatworms the adult breaks apart into two or more pieces, each of which then grows and reforms the parts it lacks to reconstitute a complete animal. Sometimes the fragments are roughly equal in size; more often, one is smaller and removes no vital organs from the "parent" animal. This process is similar to *regeneration* after loss or damage of parts of the body, which is considered at length in *Development* in this series. In other cases, the ovum does not fuse with a sperm, but it develops, nevertheless, by successive stages into a normal adult. This process, called *parthenogenesis*, is fairly widespread among invertebrates and can be brought about artificially even in mammals. In many animals where it occurs, however, it alternates with sexual reproduction from time to time, for reasons of long-term strategy of survival. Such is the case in aphids (insects), which multiply rapidly all summer by parthenogenesis but may then have a sexual phase to produce the terminal or over-wintering generation, which usually has a prolonged diapause. (See Fig. 9-1.) Parthenogenetic offspring are always of a single sex, often female but occasionally male. In the case of aphids, only females are produced by parthenogenesis and males are quite sparse except for the penultimate generation of the season. In other insects, such as honeybees and some of their relatives, unfertilized eggs develop into haploid males whereas fertilized eggs develop into diploid females.

Several species of neotropical fish are believed to be completely parthenogenetic. In a species of molly, for example, no males have ever been found and are believed not to exist. These female fish regularly mate with males of related species. The sperm, here, appear to serve only to initiate development and otherwise contribute nothing to the offspring.

Ovaries and testes are usually rather compact organs containing not only cells that divide to form the gametes but others that serve as

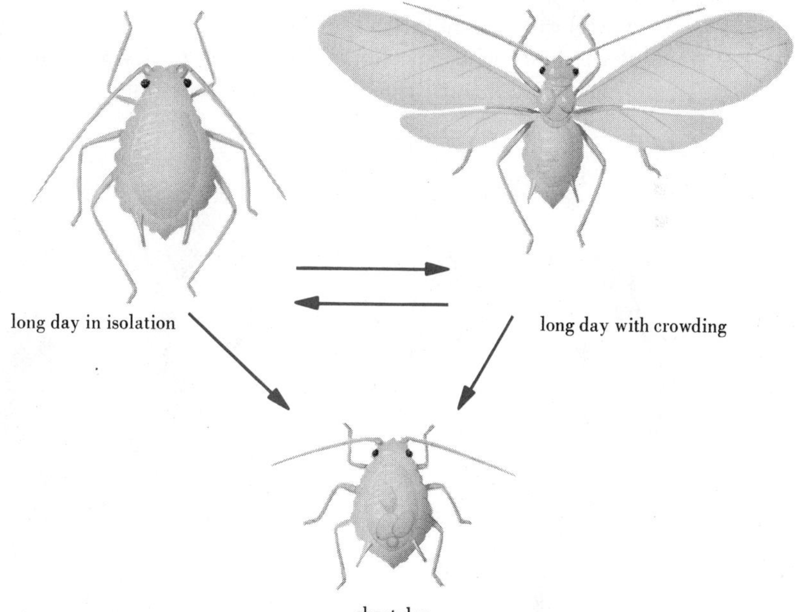


***Fig. 9-1*** *Cycle of reproduction and polymorphism in aphids. The two types of females* (above) *are parthenogenetic, producing live-born female young throughout the summer. Under crowded conditions the winged form develops; isolated individuals are often wingless. Egg-laying females, (and males) appear during the short days of autumn.*

supporting connective tissue. In some animals, certain interstitial cells of the gonads produce hormones that help to coordinate the activities of the gonads with other parts of the body.

In coelenterates, the gonads are compact masses of cells that release gametes directly into the surrounding water. In many flatworms, ovarian and testicular tissue is divided into numerous small pockets widely scattered through the body of the animal (see Fig. 2-2). Since the gonads lie relatively deep inside the animals, they can discharge the gametes (suspended in a solution similar to tissue fluids) only through a series of branching ducts leading through a genital pore to the outside. Nematode worms usually have paired ovaries or testes, or both, since many are hermaphroditic. In annelids, the gonads are paired organs in each of several segments in the central portion of the body; gametes are often discharged via the body cavity and through openings from it to the outside. Mollusks and arthropods have paired gonads with well-developed genital ducts leading to the outside. Examples of complex reproductive systems in the male and female bee are shown diagrammatically in Fig. 9-2. Insects, generally, have exceptionally complicated reproductive systems.

In vertebrate animals, the paired gonads lie against the inner surface of the body wall, just outside the body cavity and immediately under its peritoneal lining. (See Fig. 9-3.) In some but not all mammals,

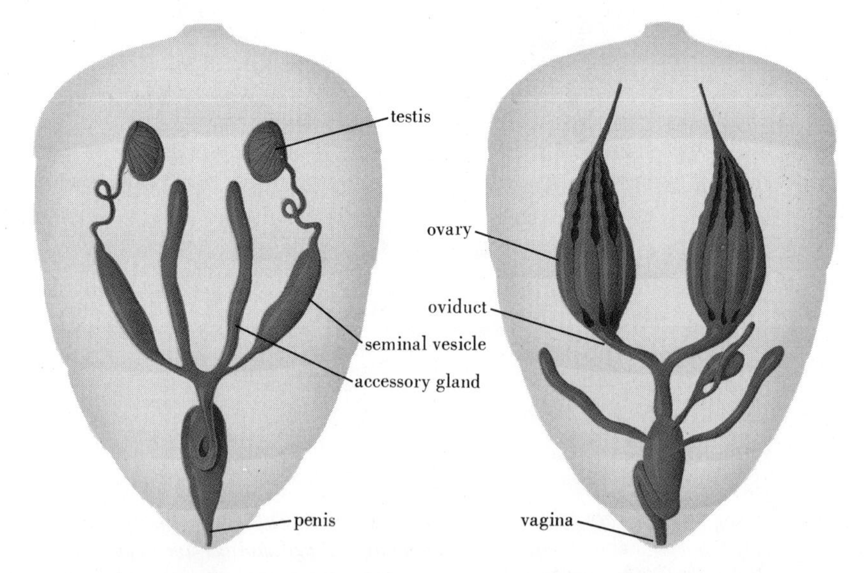


***Fig. 9-2*** *Anatomy of the reproductive system in a male* (left) *and female* (right) *bee. At the upper end of the vagina is a sperm sac which stores sperm (for months or years) received in copulation.*

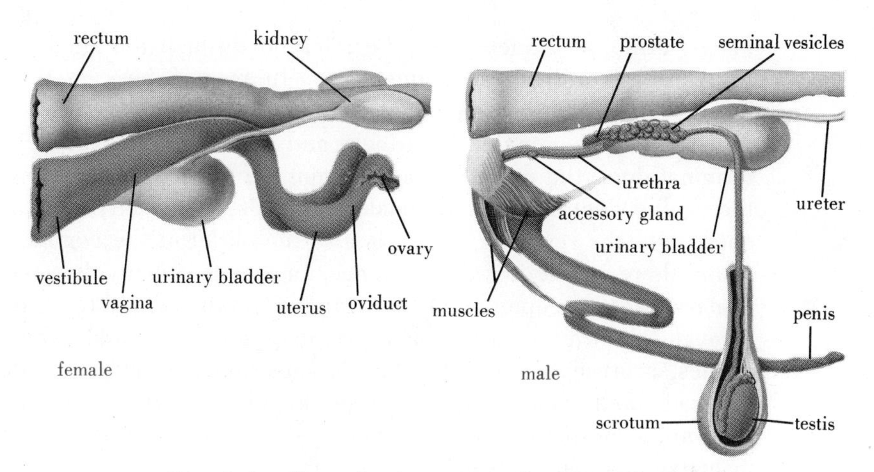


*Fig. 9-3 Reproductive system of a cow and a bull.*

the testes are contained in a pouch, the *scrotum*, which extends outside the body wall. This pouch, however, is formed by an outpocketing of the abdominal cavity and the body wall and contains a portion of the peritoneal lining in normal relationship to the testis. In many mammals, the testes descend into the scrotum only seasonally for a mating period (known, in the male, as the *rut*). In such mammals and in those with permanently scrotal testes, the production of normal sperm occurs only when the testicular tissue is several degrees cooler than the usual mammalian body temperature of about 98–102°F. In relation to this function, the scrotum is not only external but is usually relatively poorly insulated by fur. The wall of the scrotum consists of several layers of muscle fibers embedded in connective tissue. The position of the scrotum can be altered in response to external temperature and other stimuli, but the physiology and mechanisms for these responses have, as yet, been poorly worked out. Testicular cells continue to produce hormones at either body or scrotal temperature but *spermatogenesis* ceases at internal temperatures. The few mammals with permanently internal testes presumably differ in physiology from the majority, but studies to date have not revealed their novel characteristics.

The sperm of vertebrate animals are conveyed to the outside in ducts that are often shared in part with the excretory system; in most invertebrate animals the excretory ducts are quite separate from the genital openings. The evolutionary history of the association between the testis and the kidney and its ducts can be read in detail in textbooks of comparative vertebrate anatomy. In most amphibians, reptiles, and birds, the urinary and *genital ducts* and the intestinal tract open into a common but often complexly subdivided chamber called the *cloaca*.

Feces, urine, or gametes may be released through the single cloacal aperture, according to circumstances. In most fishes, on the other hand, there are distinct openings for gametes, urine, and feces.

In male mammals, the urinary and genital tracts usually share a terminal duct, the *urethra,* leading from the *urinary bladder* and from the genital duct as well. In female mammals, the urinary tract usually opens into the *vestibule,* which is also the outlet of the *vagina.* Thus, again, there is a terminal confluence. In female rodents, however, the urinary tract may open totally separately from the genital tract, passing through the *clitoris.* In the most primitive living mammals, the monotremes, represented only by the platypus and the spiny anteaters of Australia and New Guinea, a reptilian cloaca persists. Incomplete separation of these three systems is also seen in tenrecs, primitive insectivorous mammals of Madagascar.

The portion of the female reproductive tract that conveys ova from the ovary to the outside is called an *oviduct.* Often the oviduct is lined with glands that secrete a protective shell around the ovum. In animals where the young develop inside the female, the enlarged portion of the oviduct, where this occurs, is usually called the *uterus.*

## *EXTERNAL AND INTERNAL FERTILIZATION*

In most aquatic invertebrate animals, the ova and sperm are released into the water, whereupon the sperm swim to ova, perhaps attracted by special chemical substances that the ova release. Even in these animals, however, the union of gametes is not entirely a matter of chance, for the ova and sperm are often released at approximately the same time and in the same place as a result of courtship behavior carried out by the two adults, as a result of chemical *(pheromonal)* stimulation or as the result of coordinated timing tied to diurnal, lunar, or tidal rhythms. Nevertheless, the process of *external fertilization* is a relatively wasteful one; large numbers of gametes must be produced in order that a few zygotes may succeed in growing into mature animals. Any arrangement that increases the likelihood of sperm fertilizing ova and of zygotes growing into mature animals is clearly advantageous to the species since it permits a new generation to reach maturity with a smaller investment in gamete production. One of many devices evolved by various animals to meet this need is *internal fertilization.* In systems of this sort, the fertilized ova are retained inside some portion of the female reproductive tract until they complete part of their development; the sperm must, of course, reach the ovum before development can begin. Another general method for increasing the chances that a zygote will survive and grow to maturity is to have it surrounded by a protective case or *shell*; here, again, the sperm must usually reach the ovum before the shell is laid down and closes off access.

A wide variety of methods are employed to deliver sperm into the female reproductive tract. In some arthropods and mollusks, the sperm are enclosed inside a packet or *spermatophore* which is inserted into the female genital tract either by the male or by the female using one of its jointed appendages. In spiders, for example, the male, after appropriate courtship, deposits a drop of *seminal fluid* on a small silk pallet, then picks it up with his absorptive *pedi-palp* (an anterior appendage) and introduces pedi-palp and all into the female's *genital pore.* In cephalopods (squids and their relatives), the male places a cluster of spermatophores on the female's body or into her mantle cavity using a specialized arm. In one species, the specialized arm detaches from the male at the time of mating and moves about, for some time, on the female's body. Such a detached arm was once actually described as a new and separate species.

In the sharks and rays, the pelvic fins of the male are modified as grooved *claspers*. Sperm are passed from the genital pore along the groove in the clasper to the tip, which is placed against but not actually into the female genital pore. External fertilization is the rule in bony fish but in some, such as guppies, the anal fin is modified as a *gonopodium*, which functions much the same as a clasper and serves the function of internal fertilization. In many fish, elaborate courtships may insure close proximity of the male and female genital pores at the time of gamete emission (as in paradise fish or siamese fighting fish among common aquarium species) and also assure close timing of gamete emission by linking the physiological release of gametes to a sequence of mutually stimulating events. The nervous system and endocrine glands are, of course, engaged in integrating such behavior. In both kinds of fish, those with external fertilization and those with internal fertilization, ritualized courtship behavior is often required to negate the fear of one mate of the close approach of the other and may also be required to cancel normal territorial behavior. (See Fig. 9-4.) In salamanders, the male deposits a spermatophore, or packet of sperm, which is detected by the female who picks it up using the lips of her cloaca. In frogs and toads, external fertilization is the usual style but it involves behavior called *amplexus*, in which the male clasps the female and, as in fish, ensures the emission of ova and sperm into the water in a coordinated fashion and in close proximity by reciprocally stimulating behavior.

In birds, internal fertilization is universal. Normally lacking any organs for intromission, birds display elaborate courtships culminating in the approximation or pressing together of the gaping male and female

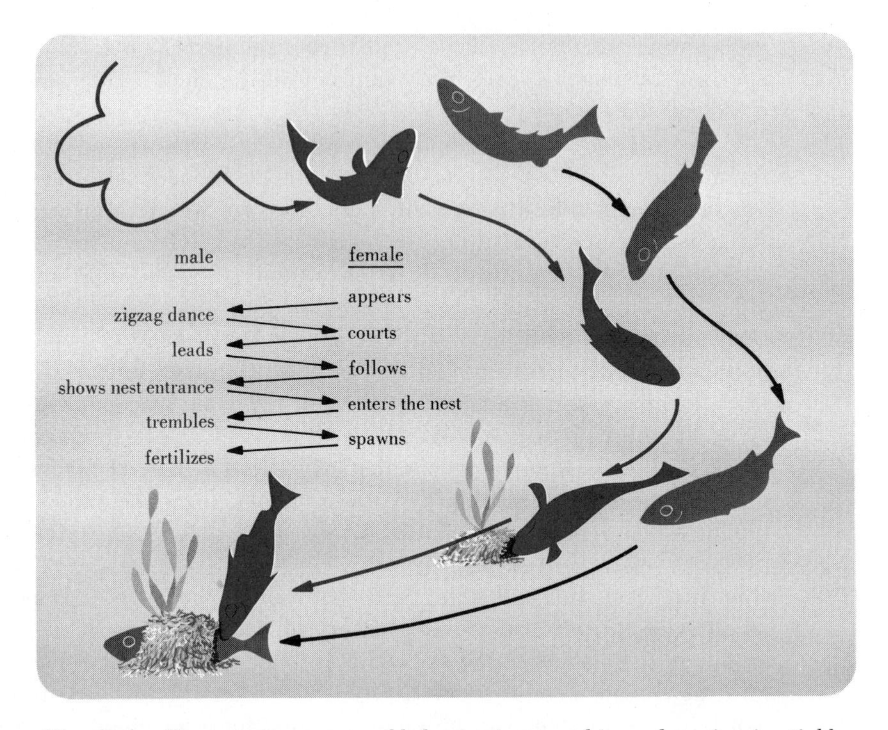


***Fig. 9-4*** *The usual sequence of behavior in courtship and mating in sticklebacks. The male builds the nest and awaits the approach of the female whom he recognizes by the full contours of her abdomen. The male's zig-zag dance is released and this in turn may release a courtship response on the part of the female. The male then "leads" the female to the nest and points to the entrance with his head. The female enters the nest, whereupon the male thrusts at her, causing her to spawn. The eggs settle in the nest and provide a chemical stimulus (a pheromone) that causes the male to eject sperm. Fertilization takes place. The sequence depends on species and sex recognition, physiological readiness to spawn, the presence of a nest, and the sequential timing and proximity (even contact) of the partners. (Adapted from Tinbergen,* The Study of Instinct, *Oxford University Press.)*

cloacas. In insects and some other invertebrates, reptiles, and mammals, the male usually has a tubular organ called a *penis*, extending the genital pore externally and designed for intromission into the female genital pore. The female genital pore is usually associated with a receptive chamber called a vagina. The penis is usually designed to be retracted in whole or in part except during mating. The vagina is usually closed except during mating and in some cases the vaginal lumen may actually be obliterated anatomically, as well as physiologically, in a cyclical fashion. The proximity and contact required for *copulation*, the time often required for emission of the sperm and, in many cases, to stimulate ovulation, and the serious vulnerability of both the male and female to injury during their linkage has led to complex courtship and mating behavior (and physiology and morphology as well) to minimize risk and to enhance efficiency. In many insects, the genitals are of such truly fantastic shape and complexity as to resemble an elegant lock and key in function. Two closely related, but different, species cannot mate successfully because the penis of one will not fit properly into the vagina of of the other, and thus intromission either remains incomplete or deposition of sperm is inhibited.

Internal fertilization is not confined to the most highly organized phyla; its occurrence is related to the need to provide a protected environment for the developing young. The tapeworms, for example, are intestinal parasites that reproduce by discharging fertilized eggs to the outside via the feces of the host animal; in their complex system of internal fertilization the males deliver sperm into a vagina by means of a penis. In this case, it is the need for very tough, indigestible shells around the eggs that makes internal fertilization necessary. Such eggs, incidentally, may alter on being ingested by the alternate host; the shell then becomes digestible and development progresses. Aquatic animals can employ external fertilization because an unprotected zygote can often survive and grow in water, especially sea water, even in the case of highly organized animals such as fish or crustaceans. External fertilization occurs in terrestrial animals only if the gametes are deposited in an aqueous medium.

## *ACCESSORY REPRODUCTIVE ORGANS*

The reproductive tract of most animals is provided with a variety of glands. Some are designed for producing shells or nutrients for the egg. Others may provide nourishment for internally developing young. Fluids are also secreted by glands to carry the spermatozoa. Taken together such fluids are called *semen*. In the cases of internal fertilization, such seminal fluids often combine with secretions of the female oviduct and provide a medium for transportation of the spermatozoa and may also provide nutrients and a suitable milieu for these gametes until fertilization occurs. In some cases, viable sperm may be stored or housed in the female genital tract, often in a specialized chamber, for days, months, or even years. Many animals mate only once in a lifetime or only over a short period of time (minutes or hours) and thereafter the female genital tract doles out a portion of the stored sperm population as needed for fertilization. Such arrangements are well known in ants and bees; moreover, sperm storage for months also occurs in many bony fish and in some hibernating bats.

Reproductive tracts are usually furnished with glands that function to lubricate during copulation, reducing mechanical blocks to intromission.

### *THE PROTECTION AND NOURISHMENT OF DEVELOPING YOUNG*

The means of protection and nourishment of young vary widely among animal groups. For example, a zygote fertilized externally usually secretes around itself some sort of membrane, often of jellylike consistency such as that surrounding a frog's egg; but no single-celled ovum or zygote can produce as heavy a shell as can be laid down around the embryo of many terrestrial animals by glands lining the oviduct. Land snails and insects produce small eggs with very tough shells that are almost impermeable to water.

Many insects lay their eggs in locations favorable to development. Some species of insects lay their eggs in burrows into which they also carry food that is later eaten by the developing young. In some cases they actually store anesthetized, living victims, so as to ensure the nutritional quality of the food available for their young. Some insects are parasitic at this stage, laying their eggs in or on other living organisms —often in other insects, but also quite often in plants or even in mammals. At the very least, eggs are likely to be laid on appropriate food plants or such other food sources as dung or carrion.

In many species, the eggs may be cared for, guarded, oxygenated, cleaned, and so forth, by one or both parents or, among social insects, by siblings. Such cases are widespread, phylogenetically speaking. Among the insects, perhaps termites, ants, and bees are the most spectacular cases. Many species of bony fish build nests, care for the eggs, and even rear their young. Often the nests are hollowed-out pockets in the sand or structures of vegetable material and sand glued together by secretions, but many species also make floating nests of bubbles or attach their eggs to plants or to rocks. Some even provide an internal nest as in the case of mouth breeders, who carry the eggs (and often the young) in their mouth, or the sea horses, in which the eggs, and later the young, are housed in the male's abdominal pouch. The best known and most elaborate nests and nesting behavior, however (excepting only some colonial insects), are found among the birds. The parents usually brood the eggs (by setting on them, turning them, and so forth) after often very complex constructions have been undertaken to hold the eggs and to protect them and the brooding parents from predators and weather. Such elaborate nesting behavior has even been exploited by parasites. In the case of some colonial insects, as well as some birds, there are parasitic species that place their own eggs in a host species' nest and have evolved physiology and behavior such that the parasitic young are brooded and reared to adulthood often in preference to, or at the expense of, the young of the host species. The cuckoos and cowbirds are especially advanced in such nest parasitism. Proper nesting behavior includes not only keeping the eggs warm or shaded, turning them, and protecting them from predation but may also include the discarding of broken or infertile eggs and the retrieval of eggs that have rolled out of the nest (only in ground nesting birds, of course). Provision must also be made for feeding the brooding parents by relief shifts, by the mate's bringing food to the nest, by prior storage of food, or some other relatively complicated behavioral pattern. Thus one finds that nesting behavior frequently rivals courtship and mating in the richness and variety of ritual displayed.

The form of the egg should not be completely overlooked. Birds' eggs, for example, may be camouflaged by being colored or speckled. In the case of some cliff-ledge nesting birds, the eggs may occur in shapes that tend to prevent them from rolling over the edge.

Zygotes of animals that lay eggs with resistant shells must develop inside these shells until they are large enough and fully enough formed to break out and fend for themselves. Their food supply during this period is the yolk contained in the cytoplasm of the ovum or other substances laid down by the female reproductive tract just prior to the secretion of the shell. The most familiar and the most highly specialized example of this type of egg occurs in birds. In the hen's egg, the actual embryo forms only a small speck on the enormous zygote, which consists mostly of stored nutrients, called *yolk*. The *white* of a bird's egg is added by the walls of the oviduct just before formation of the hard outer *shell*; it serves both as protection and later as a source of protein food.

Even greater protection than can be offered by nests has evolved in several phyla in which the developing eggs and young are retained internally and nourished by the female. Here the female's circulatory system, by special anatomical arrangements, can provision the young as effectively as her own tissues. A great variety of designs are known for such provisioning. Examples have been studied among sharks and rays, bony fish, and a few invertebrates, such as some scorpions. Quite novel solutions have evolved many times. Thus in some of the cartilaginous fish—the rays—fingerlike projections resembling intestinal villi may grow out from the lining of the mother's oviduct, enter the embryo's spiracle, and grow down into the embryo's gastrointestinal tract, where nutrient material is then secreted. In some bony fish, the young develop in the ovaries, feeding in some cases on broken-down ovarian tissue and secretions. In other *viviparous* (live-bearing) bony fish, the young in the ovaries appear to feed principally on supernumerary, smaller siblings.

Excepting only the monotremes, all mammals retain the fertilized zygotes in the uterine part of the oviduct for some time. In marsupials, the uterus often serves only to protect and not to provision the young. Such marsupial young are born at an early stage of development, just sufficient for them to find their way from the vagina to their mother's *nipple* (and often a pouch) where they are nourished for some months on *milk*. Some marsupials, however, have developed a *placenta*, whereby the mother regularly provisions the young. The placenta has reached its fullest development, however, in so-called placental mammals (all mammals except monotremes and marsupials). Here the placenta develops as a cooperative venture of the mother's uterine lining and the embryo's enveloping membranes. In the placenta, maternal and fetal blood flow past each other, separated only by thin sinus walls. (See Fig. 9-5.) Here nutrients, respiratory gases, hormones, and urea are exchanged. In fact, the placenta becomes an endocrine organ as well and may secrete many hormones affecting principally maternal but also fetal function. Even the hemoglobin of mammalian embryos is adjusted to this stage of life by having an oxygen dissociation curve differing from that of maternal hemoglobin in such a way as to facilitate the transfer of oxygen from the mother to the embryo. (See Fig. 9-6.) The embryo also has a circulatory pathway that is quite different from that of an adult. Clearly the pulmonary circuit is uncalled for, since the lungs are not ventilated; moreover, the lungs, being unexpanded, would not permit the passage of so much blood. Oxygenated blood from the placenta enters the venous system at the fetal umbilicus and feeds into the right atrium of the heart. The atrial septum in the mammalian fetus, unlike the adult, has a large *fenestra*, or opening,

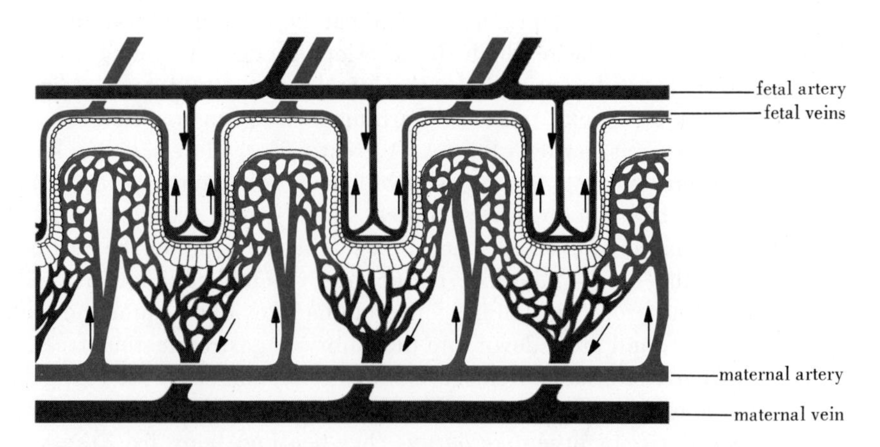


**Fig. 9-5** *The arrangement of maternal and fetal blood vessels in a sheep placenta. (After Barron.)*

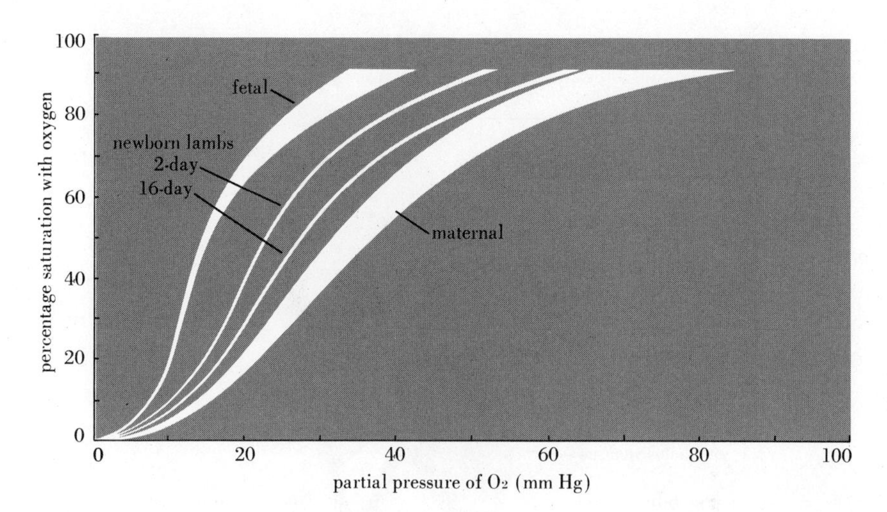


***Fig. 9-6*** *Oxygen equilibrium curves for the blood of fetal, newborn, and adult sheep. The relative positions of the curves for fetal and maternal blood indicate why oxygen will move from maternal hemoglobin to fetal hemoglobin in the placenta. At any given partial pressure of oxygen, fetal hemoglobin has a greater affinity. Even if the mother were suffering some oxygen lack, the fetus would be protected. (After Meschia et al., 1961.)*

through which much of the blood passes immediately into the left atrium. Some of the blood, however, *does* go into the right ventricle as in the adult and is pumped thence into the pulmonary artery. The pulmonary artery in the fetus is connected by a short branch, the *ductus arteriosus,* to the aorta and the remaining right-sided blood passes through this branch directly into the systemic circulation. Very little blood continues on via the pulmonary artery into the lungs. The atrial septal defect and the ductus arteriosus normally close functionally within seconds or minutes after birth and become permanently anatomically closed in several weeks by the formation of connective tissue. (See Fig. 9-7.) If one or both fetal shortcuts remain open beyond fetal life, they may seriously impair cardiac function. In man, such defects may be repaired surgically.

A great variety of other special organs play important roles in the reproduction of various animals. The abdominal appendages of many female crustaceans, for example, have provisions for attachment of the eggs, so that early stages of development can occur in this partly protected location. Certain fish have gill-like structures that are used not to gain oxygen from the water but rather to give it off to the developing eggs that are laid in water having a low partial pressure of oxygen. In pigeons, a special crop gland gives off so-called pigeon milk in both

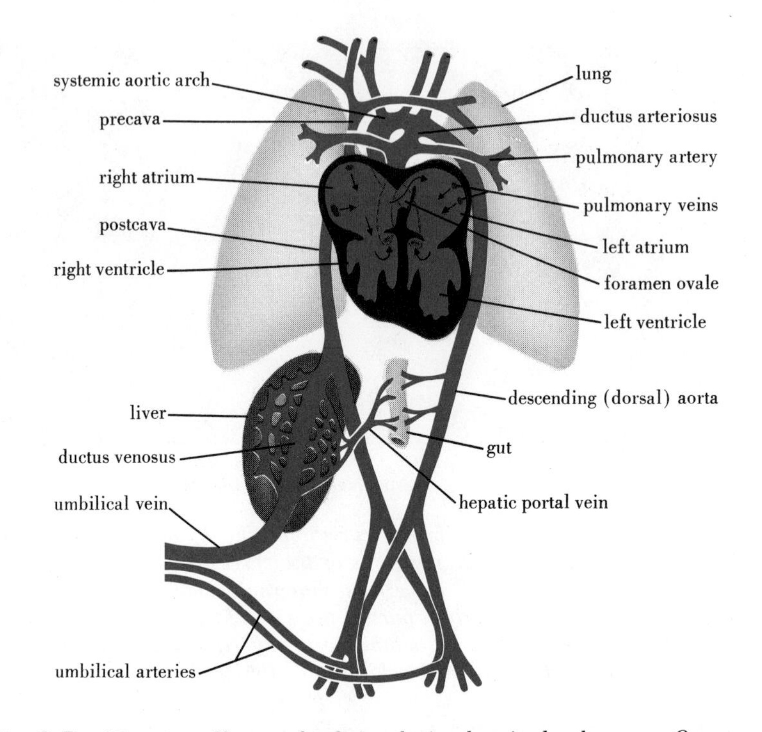


***Fig. 9-7*** *Diagram of human fetal circulation late in development. Oxygenated blood returns to the heart via the postcava. There being neither reason nor space for the blood to be circulated through the lungs, it is shunted to the left side of the heart or to the systemic circulation via the foramen ovale in the atrial septum or via the ductus arteriosus connecting the pulmonary arch with the aorta. Both these shunts normally close very soon after birth. Both may also be closed later, if necessary, by surgery.*

males and females; the young are fed on this secretion after hatching much as young mammals are fed by milk secreted by mammary glands.

These examples could be multiplied, but those already given demonstrate that many special structures have come into existence to serve the needs of reproduction along with care and nourishment of the developing young. The fact that these special organs are needed only intermittently, however, raises an important problem. A gill, a stomach, or a heart must function almost continuously, but reproduction in certain animals occurs only at certain stages of life and is often confined to certain seasons of the year. Thus the reproductive organs of most animals undergo cycles of functioning, and when not needed, they are often reduced in size. These cycles are particularly marked in the female, and perhaps most of all in female mammals, in which the uterus goes through great changes in preparation for the formation of a placenta. Animals, therefore, need not only the reproductive organs themselves, but also the means to regulate their variation in functional capacity.

The principle reason underlying seasonal reproduction seems to be to undertake this expensive metabolic activity (the production of gametes, the sustaining of the elaborate courtships, nest building, the guarding and rearing of young and, for the young themselves, growth) at a time of year favored by weather or by the abundance of food and/or water. Such factors as the absence of predators, parasites, or disease may, of course, contribute as well. In temperate climates, reproduction is often concentrated in the spring and summer. Naturally the details are dependent on *gestation period* and the period of development to adulthood, life span, and other temporal considerations. Rapidly maturing animals may reproduce many times between spring and late summer and then suspend reproductive activity for the fall and winter. In the tropics, seasonal reproduction often seems to be governed by the arrival of the wet or dry seasons – which, of course, affect food and water supplies dramatically. Frequently temperate or arctic reproductive cycles have been shown to respond to day length, a reliable physical clue to season that is available to any animal having a light-sensitive sensory system and with some opportunity for exposure to daylight. In many birds, for example, not only the development of the gonads is triggered by day length (mediated by the retina of the eye via brain pathways to the hypothalamus and thence to the anterior pituitary gland), but seasonal migrations, molting of feathers, and several other major annual events are equally so controlled. Such triggering by day length or the arrival of heavy rains may synchronize the species population so as to provide for the simultaneous readiness of males and females to mate and allow for the economy of nonreadiness at other times of year. Additional fine-grain synchronization, however, is also known to occur, especially as studied in colonial birds, in which the sight and sound of other courting individuals has been shown to enhance the reproductive readiness of all. Synchronization occurs to lunar and tidal cycles as well. Thus a fish living on the American West Coast spawns on the beaches (leaping out onto the sand above the high-water mark) on several nights following the highest tide of a given lunar cycle. The eggs develop in the sand and the embryos hatch almost instantly when the next sufficiently high tide engulfs them, some 25 or 27 days later. Reproductive cycles may also have shorter periods than a year. Some parasites – especially those carrying malaria, for example – may cycle every few days; the triggers are not usually known. In mammals, the female of almost all species has a cycle of receptiveness or readiness to mate called *estrus*. This is often triggered by day length and involves the ovarian production of estrogens and progesterones and the maturation of ova. Usually, at the peak of her receptiveness, the female has either recently ovulated or is ready to ovulate. Thus sperm are likely to be available for fertilization while the egg is fresh. Estrus cycles may be spaced several weeks apart, as is the case in many small rodents, or two or more years apart, as in very large mammals such as whales or elephants. The cycles are suspended during pregnancy. The males of many mammalian species are perpetually prepared to mate productively but, in others, the testes undertake spermatogenesis only seasonally, at a time often associated with the seasonal descent of the testes into scrotal position. Such a seasonal rutting period often coincides with the production of sexually significant, odoriferous, glandular products that may attract females, and with special behavior related to the seeking out and courting of receptive females.

Many other physiological and behavioral arrangements have evolved for increasing the efficiency of reproduction. These arrangements include a rich variety of devices for attracting or courting potential mates. Sexually attractive odors are widely used in terrestrial groups such as mammals and insects. The sex attractant of female gypsy moths has been said to be effective at a distance of 2 miles. The male moth flies upwind until he encounters the source of the odor. Male moths can be collected in surprising numbers by exploiting a sexually attractive female. The appeasement of hostile behavior, which might otherwise preclude mating, often involves eleborate ritualized posturing, the display of special visual patterns, the use of tactile stimuli, and so on. Courtship, in addition, may be based on or include long-term bonds between the pair, which facilitate not only mating but the rearing of young and the division of labor as well. Such pair bonds occur among some birds and mammals. In other species, herds or colonies exist that may be based on one or more dominant males. Many other colonial and social arrangements, often associated with reproduction, are also known. Some additional aspects of reproduction are dealt with in *Behavior* in this series.

## FURTHER READING

Amoroso, E. C., "Viviparity in Fishes," *Symposium of the Zoological Society of London,* 1:153–181, 1960.

Bastock, M., *Courtship; An Ethological Study.* Chicago: Aldine, 1967.

Beach, F. A. (ed.), *Sex and Behavior.* New York: Wiley, 1965.

Benoit, J., "Hypothalamo-Hypophyseal Control of the Sexual Activity in Birds," *General and Comparative Endocrinology (Supplement),* 1: 254–274, 1962.

Breder, C. M., and D. E. Rosen, *Modes of Reproduction in Fishes.* New York: Natural History Press, 1968.

Bullough, W. S., *Vertebrate Sexual Cycles*. London: Methuen, 1961.

Carr, A., *So Excellent a Fishe*. New York: Natural History Press, 1967.

Etkin, W. (ed.), *Social Behavior and Organization Among Vertebrates*. Chicago: University of Chicago Press, 1964.

Farner, D. S., "The Photoperiodic Control of Reproductive Cycles in Birds," *American Scientist*. 52:137–156, 1964.

Galtsoff, P. S., "Physiology of Reproduction in Molluscs," *American Zoologist*, 1:273–289, 1961.

Hoar, W., *General and Comparative Physiology*. Englewood Cliffs, N. J.: Prentice-Hall, 1966.

Jones, J. C., "The Sexual Life of the Mosquito," *Scientific American*, 218 (4): 108–116, 1968.

Loewy, A. G., and P. Siekevitz, *Cell Structure and Function*, 2d ed. New York: Holt, Rinehart and Winston, 1969.

Parkes, A. S. (ed.), *Physiology of Reproduction*, 3d ed. London: Longmans, 1958–60.

Rheingold, H. L., *Maternal Behavior in Mammals*. New York: Wiley, 1963.

Romer, A. S., *The Vertebrate Body*, 3d ed. Philadelphia: Saunders, 1962.

Rothschild, M., and T. Clay, *Fleas, Flukes and Cuckoos*. New York: Collins, 1952.

Tienhoven, A. von, *Reproductive Physiology of Vertebrates*. Philadelphia: Saunders, 1968.

Wigglesworth, V. B., "The Hormonal Regulation of Growth and Reproduction in Insects," *Advances in Insect Physiology*, 2:247-336.

Young, W. C. (ed.), *Sex and Internal Secretions*, 3d ed. vols. 1 and 2. Baltimore: Williams & Wilkins, 1961.

# chapter 10

# Coordination of Function

If an animal's muscles contracted and relaxed in a random fashion while the arterioles shunted blood here and there and the heart beat faster or more slowly, all without regard to other events in the animal's body, death would soon result. Fortunately, such anarchy is prevented by harmony of action, thanks to special organ systems that control body functions. Best known of these is the *nervous system*, but this is aided by, and intimately interrelated with, a system of *endocrine glands* that regulate the growth or activities of other tissues.

***THE STRUCTURE OF NERVOUS SYSTEMS*** All multicellular animals (except sponges) contain long nerve cells, called *neurons*, portions of which, the *axons*, are collected together into parallel bundles called *nerve trunks.* Any nerve trunk large enough to be visible with the unaided eye contains hundreds or sometimes thousands of axons. These control the contractions of muscles and the actions of other effectors as well as the conduction of information from sense organs to the rest of the animal. After their nerve trunks have been cut, muscles remain toneless or flaccid, or the animal ceases to react to stimulation of the corresponding sense organs. Even in some of the ciliates there are *intracellular fibrils,* which run just below the cell membrane and interconnect the *basal bodies* of cilia or groups of cilia. When they are cut, the beating of the several cilia loses its coordinated timing and is much less effective in moving the animal through the water. Thus these fibrils, though not neurons, serve a function something like that of neurons. Coelenterates achieve coordination of their body movements by a network of neurons; this network, often called a nerve net, consists of distinct cells. The neurons are especially concentrated around the mouth in some coelenterates, but the nerve net ramifies, as well, throughout the body.

Neurons have a cell body containing a nucleus and cytoplasm much like that of other cells; from the cell body one or more highly elongated processes radiate outward. One type, the axon, usually extends for some distance. (See Fig. 10-1.) Axons, which make up the bulk of the visible nerve trunks, frequently run parallel to one another and each may be surrounded along its length by a series of *sheath cells* (often, in vertebrates, called Schwann cells). During embryonic development, in vertebrates, the axon fits into an indentation in the sheath cell. The fully developed axon looks as if it had rotated on its axis during the growth of the cells so that many layers of sheath-cell membrane had come to envelop the axon like a "jelly roll." No actual rotation has been observed. The mechanism for the formation of the "jelly roll" is not currently known. The compound membrane, called the *myelin sheath*, contains alternate layers of fat and protein molecules, and when sufficiently thick gives a group of axons a whitish appearance. (See Fig. 10-1.) Even within any one nerve trunk, the axons vary greatly in size and in thickness of the myelin sheath; the smallest are almost invisible in the light microscope, having diameters of less than 0.2 $\mu$ and only a single layer of sheath-cell membrane. The largest neurons (unmyelinated), found in some of the cephalopods and annelids, are a few hundred microns in diameter. In vertebrate animals, the larger axons have thicker myelin sheaths; the axon proper may be as much as 50 $\mu$ in diameter. Some axons are extraordinarily long; those running from the spinal cord to the digits of large mammals may extend 2 meters or more. While axons ordinarily remain about the same diameter along most of their length, they often branch near the ends, and the individual branches are usually smaller in diameter than the axon that gave rise to them.

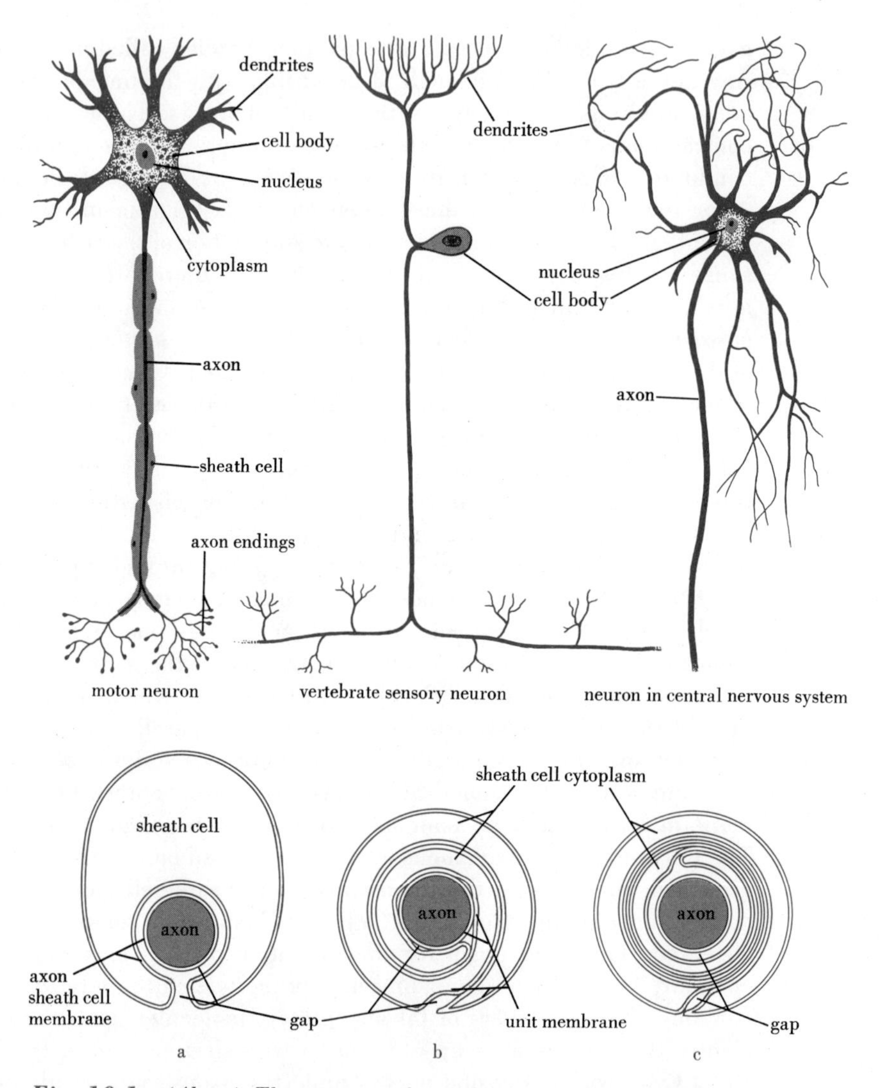


***Fig. 10-1*** *(Above) Three types of vertebrate neurons. Almost all neurons are much longer, relative to the diameter of the axon and cell body, and many have much thinner myelin sheaths than those shown here.* (Below) *The sheath cells initially surround an axon in a simple geometry. As development proceeds, however, the sheath cell multiply enfolds the axon, finally making up the mature myelin sheath of peripheral nerves. (After J. D. Robertson in* Biophysics of Physiological and Pharmacological Actions. *A. M. Shanes, Ed., Amer. Assoc. Adv. Sci., 1961.)*

In addition to axons, the cell bodies of many neurons give off smaller processes, called *dendrites* because they are usually highly branched like trees. Nerve-cell bodies and their dendrites are usually aggregated into concentrated masses of nervous tissue called *ganglia* (singular ganglion); an animal's main assemblage of ganglia is called its *central nervous system*. Within ganglia, the ends of axons come very close to the cell bodies and/or dendrites of other neurons; these places of contact, called *synapses*, are of the greatest importance in the functioning of nervous systems. At a synapse, the two cell membranes are usually somewhat thickened; the gap between the two may be as small as 200 Å and hence only visible by means of the electron microscope. A specialized type of synapse, called a *neuromuscular junction*, occurs where the axon of a motor nerve ends on a muscle cell.

Within the central nervous systems of vertebrate animals, one finds bundles of axons with myelin sheaths much like nerve trunks found elsewhere in the body. These are often sufficiently concentrated to give a whitish appearance to the tissue. Such parts of central nervous systems are called "white matter," to distinguish them from "gray matter," which contains predominantly cell bodies and dendrites. The color reflects the presence or absence of myelin in quantity. In vertebrates, these ganglia of the central nervous system are called *nuclei*. Ganglia and central nervous systems in general also contain large numbers of cells, called *glia*, that bear a superficial resemblance to connective tissue cells. Even though glial cells may make up a major part of the bulk of a vertebrate central nervous system nucleus, for example, their functions have not been established. Their relationship to neurons is so close, however, that they may well be the counterparts, in the central nervous system, of the sheath cells that lie on the axons of peripheral neurons, serving, in part, an insulating function. It has been suggested recently that glial cells may participate actively in the storage of information or in the organization of such storage in the central nervous system.

In most flatworms and roundworms, the nervous system consists of two main nerve trunks, with numerous cross connections, interspersed with ganglia. They have slightly larger ganglia in the head region. This tendency to concentrate central nervous system tissue into the anterior end is more pronounced in annelids and arthropods, which also have a double nerve cord running the length of the animal with cross-connecting nerve trunks in each segment. This double nerve cord is located ventral to the digestive tract. The main, anterior part of the central nervous system, which in these animals is prominent enough to be called a *brain*, consists of ganglia lying both ventral and dorsal to the digestive tube with lateral connectives forming a ring around the esophagus. (See Fig. 10-2 & 10-3.) The concentration of ganglia into the anterior end is related to the corresponding tendency toward concentration there of the major sense organs. These include eyes and sensory

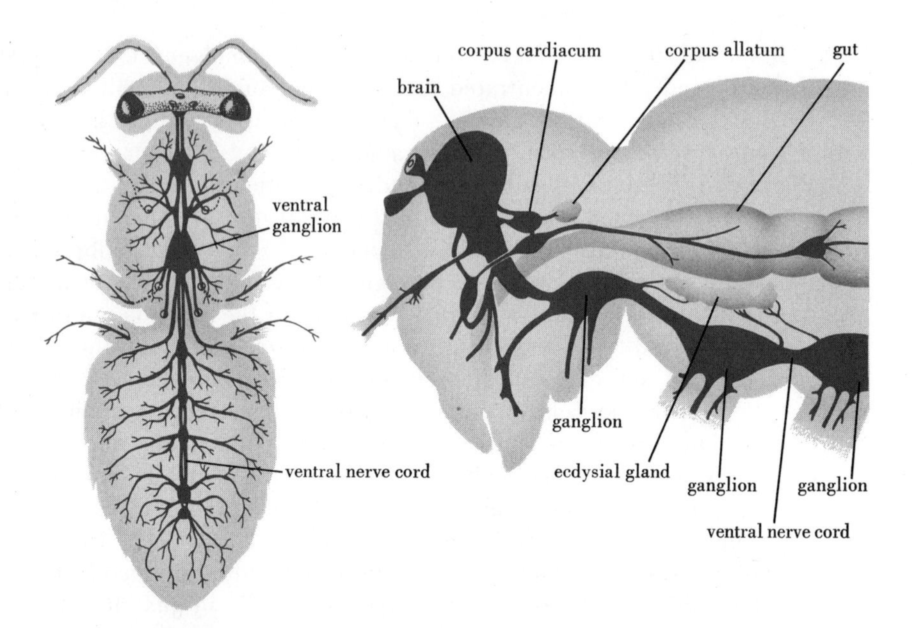


***Fig. 10-2*** *A diagram of the nervous system of the honeybee* (left). *The ganglia and nerve trunks shown represent aggregates of hundreds or thousands of neurons. (After R. E. Snodgrass,* Anatomy of the Honeybee, *Cromstock Press, 1956.) The anterior end of a generalized insect* (right) *shows the elaborate development of the nervous system in the head and surrounding the esophagus as well as the close association between the nervous system and two endocrine organs, the corpus allatum and the ecdysial gland. (After Weber, 1952.)*

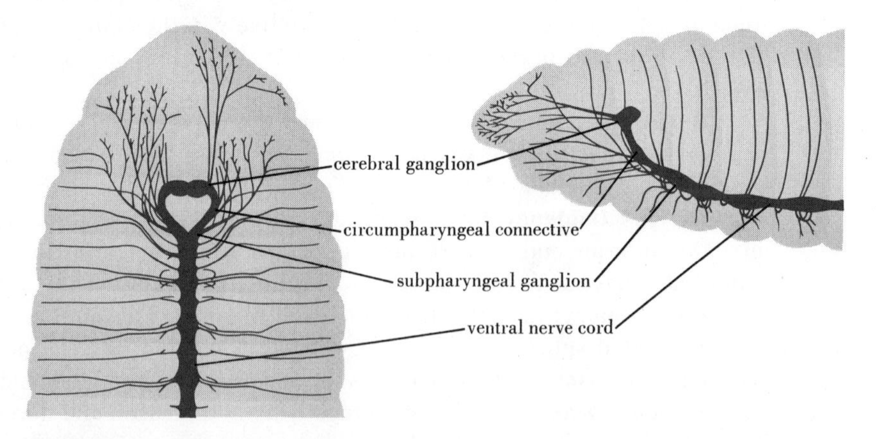


***Fig. 10-3*** *The nerves of the anterior end of an annelid, the earthworm,* Lumbricus. *As in insects, the nervous system is well developed anteriorly and consists of a series of ganglia connected by a ventral nerve cord. (After T. H. Bullock and G. A. Horridge,* Structure and Function in the Nervous System of Invertebrates, *Freeman, 1965.)*

cells close to the mouth that are stimulated mechanically and chemically by food and other substances. In mollusks, the central nervous system has no simple structural pattern, but there are sizable ganglia near all of the major organs, such as the digestive tract and the foot. In the cephalopods, there is a brain near the well-developed eyes and mouth—which, like that of the annelids and arthropods, takes the form of a ring surrounding the esophagus.

In vertebrate animals, the central nervous system is constructed on quite a different plan. There is a single nerve cord dorsal to the digestive tract and hollow along its whole length. The anterior part of the central nervous system, lying in the head, is called the *brain* and is continuous with the *spinal cord* in the neck, trunk, and tail. (See Fig. 10-5.) The axial skeleton, the chain of vertebrae and the skull, surrounds and protects the central nervous system and the major cranial sense organs. The *spinal canal* in the center of the spinal cord is continuous with the *ventricles* of the brain; all are filled with *cerebrospinal fluid*, similar to lymph, which is secreted by specialized capillary systems lying in the walls of several of the brain ventricles. The body of a vertebrate animal is segmented in some ways, though less obviously so than that of annelids and arthropods. This segmentation is reflected in the emergence from the spinal cord of two *spinal nerves*, one on each side, from each segment, through notches in the side of each vertebra. These spinal nerves consist, in part, of axons of motor nerves running from their cell bodies in the gray matter of the ventral portion of the spinal cord of that segment to striated muscle cells. In addition, axons of sensory neurons run from peripheral sense organs—such as those of touch, temperature, or pain—in or near the skin to their cell bodies, which lie in *sensory ganglia* just outside the spinal cord, one on each side in each segment. There are also axons of motor cells of the so-called *autonomic nervous system*, which derive from cell bodies in the ventral horn of the spinal cord and ultimately innervate smooth muscle cells and glands (but only after a synapse with a second neuron). In addition to the brain and spinal cord, there is in vertebrates a chain of *sympathetic ganglia* (some paired, others not) lying lateral and ventral to the spinal cord in the neck and trunk. In these ganglia, some of the axons of the autonomic nervous system, the *sympathetic* portion, synapse with a second axon which then innervates smooth muscle cells (or a gland) such as those of the gut or of blood-vessel walls. There are also smaller but much more numerous ganglia, called *parasympathetic*, distributed in the walls of many of the viscera or in their near vicinity. There, the remaining motor axons of the autonomic nervous system, emerging only from the cranial and sacral parts of the central nervous system, synapse with a second neuron destined also to innervate a smooth muscle cell or a gland.

From the brain arise paired *cranial nerves*, which differ greatly in size and importance. Fundamentally they have the same components as spinal nerves, but often these are more specialized in association with the development of major senses and special motor functions in the anterior or head end of the body. The segmented arrangement of the cranial nerves is often more difficult to perceive because of the complexity of development that the brain exhibits and because of the long and complex evolutionary alterations in skull makeup that have occurred. In mammals, custom has defined 12 pairs of cranial nerves, although some additional ones are known in lower vertebrates; some are clearly fused pairs or groups of cranial nerves, and one, the hypoglossal, can really best be called the most anterior of the spinal nerves. (See Fig. 10-4.) In mammalian anatomy, however, and, indeed, in much of vertebrate anatomy, convention is strong and nomenclature often yields to such convention. The first of the cranial nerves, the *olfactory nerves*, are concerned with the special sense of smell, an important one in vertebrates and, perhaps, historically the prime sense. The second pair, the *optic nerves*, are also sensory (and are really brain tracts, since the retina, itself, is derived from the brain), conveying information from the retina to the brain. The third, fourth, and sixth pairs of cranial nerves are simply motor nerves controlling the action of the striated muscles moving the eyeballs. The fifth, seventh, ninth, tenth, and eleventh nerves are called *branchial* and are all of like design. Historically these nerves were associated with the gill arches. The skeletal and muscular structure of the gill arches has undergone, in higher vertebrates, almost incredible evolutionary modification into a very large number of seemingly unrelated structures. These structures, however, continue to possess their primitive branchial (or gill) innervation. Thus, in mammals, the fifth cranial nerve, the *trigeminal*, which innervates the jaws and the muscles of chewing, is derived from a gill arch anterior

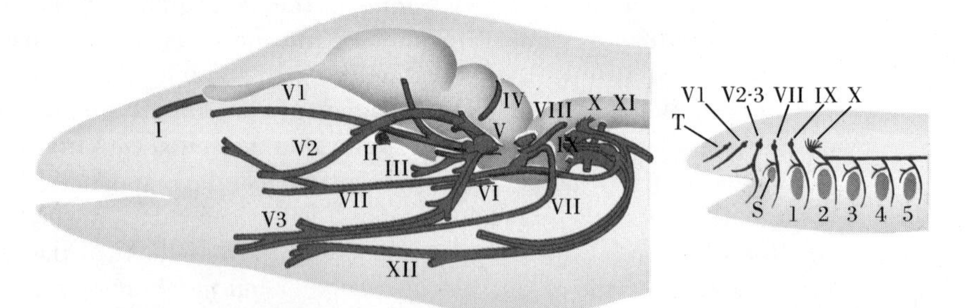


***Fig. 10-4*** (Left) *Diagram of the distribution and components of the cranial nerves (I to XII) of a lizard.* (Right) *Diagram showing the distribution of branchial cranial nerves (those originally associated with gill slits) in a jaw-bearing fish.* T *indicates the terminal nerve (often lost);* S *indicates the spiracle. The arabic numbers indicate the typical gill slits of fish. (After Romer, A. S.,* The Vertebrate Body, *3rd ed., Saunders.)*

to the spiracle. It also innervates a small muscle of the middle ear, which tenses the eardrum and alters its responsiveness to sound. The seventh cranial nerve, the *facial*, innervates the muscles of facial expression in man, those with which we alter the shape or posture of our lips, nose, eyelids, and cheeks. This nerve originally supplied the spiracle. It also innervates another small middle ear muscle, the *stapedius*, which controls the posture of the *stapes*, one of the auditory ossicles. This also protects the ear from loud sounds. The history of the middle ear and other derivatives of the branchial region is dealt with in textbooks of comparative anatomy. The ninth cranial nerve, the *glossopharyngeal*, originally associated with the first gill arch behind the spiracle, innervates certain small muscles of the palate and pharynx. The tenth cranial nerve, the *vagus*, acts with the eleventh, the *accessory*, to innervate all of those structures derived from gill arches behind the glossopharyngeal arch, which include a large part of the pharynx, the larynx, and the trachea. Of these branchial nerves, the seventh, ninth, and tenth also carry axons of the special cranial sense of *taste*. Through these branchial nerves as well, the cranial part of the *parasympathetic nervous system* flows to ganglia in salivary glands, the respiratory tree, the anterior end of the gut, and the heart. Most of this parasympathetic outflow is via the vagus nerve.

The eighth cranial nerve, the *auditory*, conveys sensory axons, from the inner ear, concerned with hearing and detection of gravity and linear and angular acceleration, to the brain. In fish, this cranial nerve is, perhaps, even more important than in mammals in that it is associated with the *lateral line system* of sense organs distributed over the surface of the head and body. The lateral line organs appear to measure movements and water currents in their vicinity. The twelfth cranial nerve, the *hypoglossal*, innervates the muscles of the tongue and certain striated muscles in the neck. Primitively, this nerve was spinal and its structure and function are essentially spinal in mammals.

In vertebrates, information feeds into the central nervous system from sense organs, which may be in or near the skin, monitoring temperature, pain, touch, or pressure, or may be in muscles or joints, measuring muscle or tendon tension and joint position. Information also comes from more localized and specialized sense organs for chemical senses (taste and smell), vision, hearing, and the measuring of gravity and acceleration. Internal sensory systems also monitor body temperature, carbon dioxide tension in the blood, blood osmotic pressure, and blood pressure. Information about the environment (external or internal) is coded by the sense organs and relayed to the central nervous system, the brain and spinal cord, by sensory nerves, which normally have a long axon extending centrally to a cell body lying immediately adjacent to the spinal cord or brain in a sensory or dorsal root ganglion. These sensory axons also extend from the sensory cell body into the central nervous system via the dorsal roots. Once in the central nervous system, they usually synapse with one or more *association neurons*, which distribute the acquired information to one or more further destinations. In the simplest case, the association neuron may deliver its message to a motor nerve in the ventral portion of the cord or brain stem and cause a response such as a muscle contraction or a glandular secretion. Much more often, the sensory information is dispersed to a large number of other centers or nuclei on the same and on the opposite side of the brain or spinal cord. (The central nervous system is anatomically symmetrical in almost all respects in vertebrates. The actual physiological processing may differ from side to side.) At each synapse, the information can be processed further by combinations of stimulation and inhibition, temporal and spatial summation (all of which we will discuss later in this chapter), and undoubtedly many other interactions not fully understood. Eventually motor activity may be effected. Motor or effector activity usually occurs as muscular contraction (producing locomotion, postural adjustment, or peristalsis, for example) but may also be in the form of secretory activity by glands or of *neurosecretion* by neurons or modified neurons such as occur in the hypothalamus of the brain and in the medulla of the adrenal gland. Also known are "motor" nerves, which run to some of the sense organs—the inner ear, for example—and apparently alter the sensory threshold.

The most anterior part of the brain is primitively associated with *olfaction* and continues to be so (See Fig. 10-5.), although the olfactory processing centers come to be overshadowed in both birds and mammals by the enormous development of other parts of the *forebrain*. In birds, a group of association nuclei called the *basal ganglia* are strikingly developed. These are thought to be concerned with the exceptionally elaborate stereotyped behavior that birds display. In mammals, the *cerebral hemispheres* are dominated by the development of the *cerebral cortex*, consisting of large numbers of nuclei concerned with elaborate motor behavior such as posture, locomotion, and speech, and also with complex analysis of sensory information and the interaction of such information with experience, judgment, esthetic taste, and many "higher" functions often grouped, in part, as intelligence.

Another exceptionally important part of the forebrain in all vertebrates is the *hypothalamus*, an aggregation of nuclei that process sensory information chiefly concerned with vegetative functions. Hypothalamic nuclei regulate eating and drinking, body temperature, and some aspects of reproductive behavior, the production of urine, and many other less well-known functions. The hypothalamus acts in many ways: (1) by neural output that creates, affects, or inhibits such drives as hunger and thirst; (2) by neural output that alters the circulatory

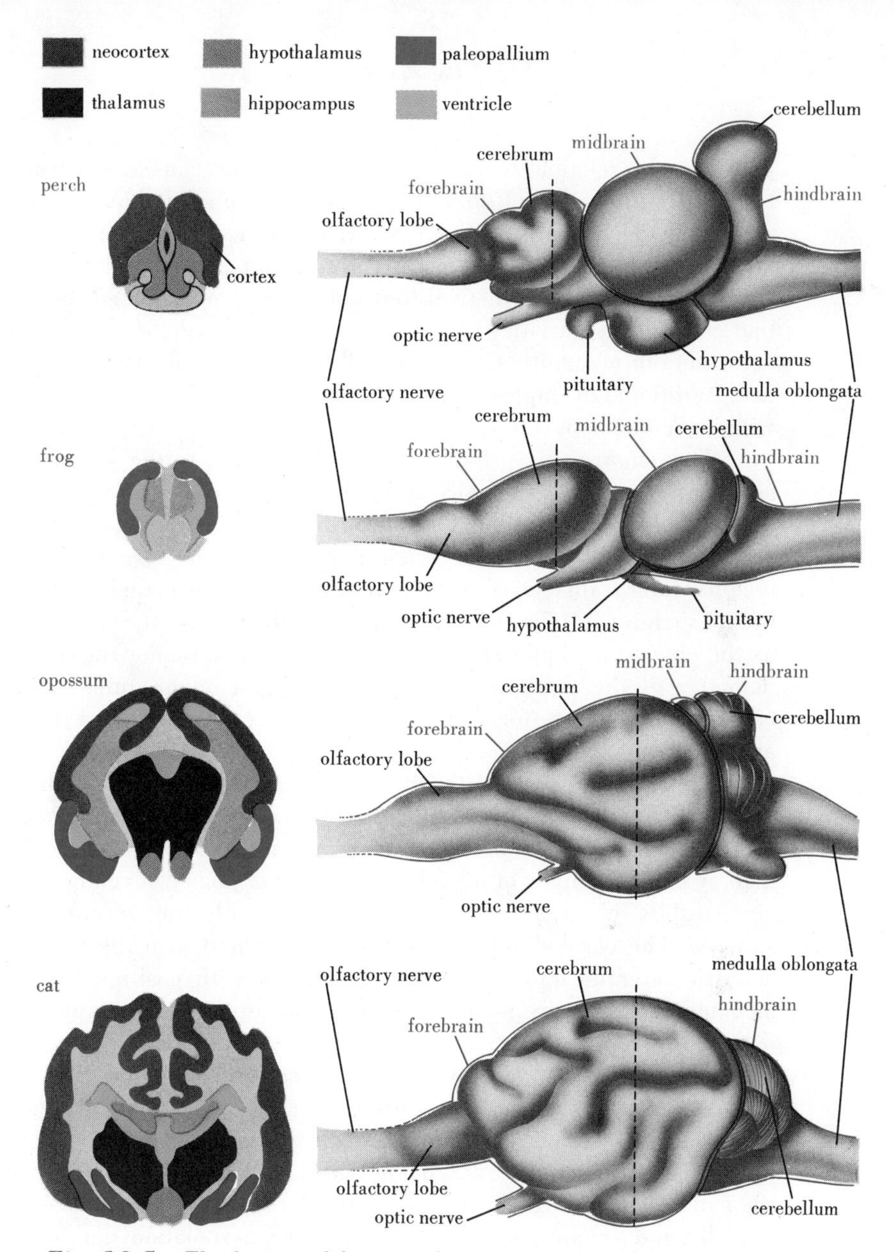


***Fig. 10-5*** *The brains of four vertebrates, a perch, a frog, an opossom, and a cat. The homologies of the parts of the cerebral cortex have not been worked out for fish and become successively more complex in tetrapods. To the older paleopallium and hippocampus of frogs is added the neocortex in mammals. In fish, amphibians, and reptiles, the midbrain (associated with vision primarily) is prominent and the forebrain is largely devoted to the handling of olfactory input. In mammals, however, the forebrain undergoes tremendous development and overshadows (even obscures from view) the midbrain. The central canal or ventricle of the brain is simple in appearance in fish but becomes subdivided in tetrapods. Its shape, as seen in cross section, depends upon the level at which the section is taken. (After W. G. Van der Kloot,* Behavior, *Holt, Rinehart and Winston, 1968.)*

pathways in relation to temperature control or causes sweating; (3) by neural output that alters behavior so that the animal moves into the sun or shade or curls up tightly; (4) or by neurosecretory output of many different products, including *antidiuretic hormone* and *oxytocin* as well as several hypothetical (but indirectly demonstrated) hormones that control the activities of the *anterior pituitary gland.* The hypothalamus acts in many other ways as well, but this small sample must suffice. Additional examples of neural and neurosecretory control are dealt with in *Behavior* in this series.

The *midbrain* is primitively largely devoted to the processing of *visual information*, especially in a complex pair of dorsal nuclei called the *optic lobes.* These lobes are less dramatic in size and complexity in mammals than in other vertebrates, since some visual processing is apparently transferred to the cerebral cortex. Nevertheless, the midbrain nuclei continue to function crucially in visual reflexes (such as the change in shape of the lens for accommodation or the change in diameter of the pupil with light), which must occur quickly, before conscious decisions (for which there is really no need) and reactions are possible.

The *hindbrain*, along with other functions summarized later, is highly associated with the *acousticolateralis nerves* (mediating *hearing, gravity,* and *acceleration senses* as well as the *lateral-line senses* of fish). A super complex of nuclei, the *cerebellum*, has evolved here, often comparable, in organizational complexity, with the cerebral hemispheres. The cerebellum is apparently concerned with the analysis of acousticolateralis information as it interacts with position sense and locomotion. Certainly some aspects of posture and coordination are determined here.

The hindbrain also includes the *medulla oblongata*, which, together with a comparable extension into the midbrain, is organized much like the spinal cord and is called the *brain stem.* Here sensory information from the head and pharynx is fed in and motor activity in the same regions is governed. As in the cord, the sensory nerves enter dorsally and synapse with dorsally grouped association neurons. The motor neurons are ventral. Large bundles of intercommunicating axons pass up and down the brain stem to higher and lower centers and between closely located nuclei as well. The brain stem is also the site of origin of the parasympathetic neurons and is the site of processing of taste information.

## *THE FUNCTION OF NERVE CELLS*

Nerve impulses, the messages that regulate the actions of other cells at axonic endings, travel along all axons. In all types of axons, the nerve impulses are basically the same, differing only in minor details. They travel rapidly, although not nearly so fast as electric current, light waves, or even sound waves. Large-diameter axons transmit impulses faster than small ones, and heavily myelinated axons transmit faster than poorly myelinated ones. Their rate of transmission increases with temperature. In mammalian axons, the rate varies from about 1 to 100 meters per second. The impulse itself is a transient set of biochemical and biophysical reactions occurring mainly at the cell membrane of the axon and accompanied by a small and very brief change in electric potential known as the *action potential*. Although the metabolic rate increases during and after passage of an impulse, the action potential, itself, is the most easily measured sign that a nerve impulse is occurring. The cell membrane of an inactive axon, like that of most other cells, has a slight disproportion of positively charged ions at its outer surface; hence the outside of the axon is electrically positive relative to the cytoplasm. Furthermore, there is a relative accumulation of potassium ions and a shortage of sodium ions inside the cell. When an impulse passes a given part of the axon, the cell membrane suddenly becomes much more permeable to positively charged ions than when it was at rest. Sodium ions then diffuse inward and potassium ions outward during the very small fraction of a second before the membrane recovers its resting properties. By active transport, the membrane continually maintains the distribution of sodium and potassium ions. The movement of these ions produces, in turn, the action potential, which is a brief change of potential in a negative direction when measured by an electrode just outside the axon. The duration of these changes at a given point on an axon is only about 1 millisecond but, because of the speed of impulse conduction, the length of axon involved at any one time is a few millimeters. So few ions flow during an impulse that many impulses can pass before the gross ion concentration changes appreciably. Nevertheless, the crucial point is that there is a local flow of ions across the membrane.

Nerve impulses are discrete, unitary events, and in a given axon, over a short period of time, they are all alike. (See Fig. 10-6.) Over longer periods, the state of an axon shifts sufficiently that nerve impulses may vary. By analogy to computing machines, the *all-or-none* nature of the nerve impulse, discovered about 50 years ago by biologists, has been transposed into the language of electrical engineers by describing conduction of impulses along axons as "digital." The axon either conducts an impulse or does not. There are no shades in between. But a given axon may vary greatly in the frequency at which nerve impulses are transmitted, from none at all or a few per second up to maximum rates of almost a thousand per second. The frequency of impulses delivered by the axon at its terminal synapses indicates the strength of excitation. For example, the higher the frequency of impulses conducted along a motor neuron, the stronger is the contraction of the

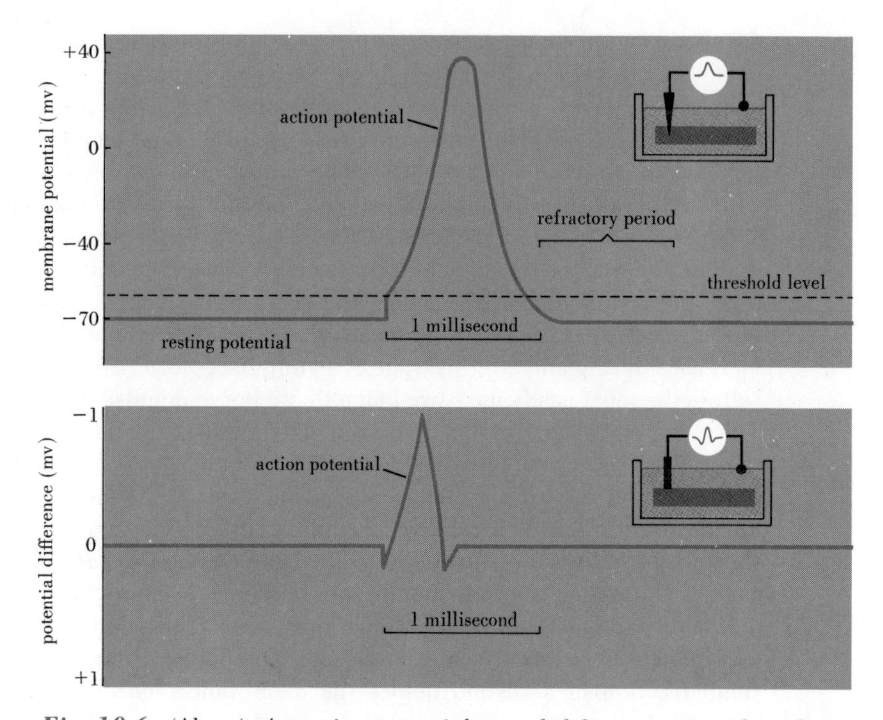


***Fig. 10-6*** (Above) *An action potential recorded by measuring the voltage difference between a microelectrode inserted into the interior of an axon (piercing the cell membrane) and a large reference electrode in the surrounding medium.* (Below) *An action potential recorded by a fine electrode, resting on the surface of an axon, and a reference electrode. Note that the shape and size of the recorded action potentials are quite different. During the refractory period the neuron at first will not respond to any stimulus, however great. Later, a response will occur but only to a strong stimulus. By the end of the refractory period (the length of which varies from cell to cell) the threshold has returned to "normal". (After W. G. Van der Kloot,* Behavior, *Holt, Rinehart and Winston, 1968.)*

muscle fibers it serves. Once this frequency has exceeded a few impulses per second, the muscle is excited into a sustained contraction as described in Chapter 6 and the strength of the contraction increases with the frequency of impulses in the motor nerve. Likewise, strong excitation of a sense organ causes a higher frequency of impulses in the axons of its sensory nerve. A general principle that applies to all parts of all nervous systems is this: whenever an axon conducts impulses at a higher frequency it is signaling a higher level of excitation, and the synapses at which it ends are influenced more strongly. This principle, long known to biologists, has also been glamorized recently by translation into the language of engineering as "the frequency modulation of axonal conduction." (Many sensory nerves have spontaneous discharge rates, which can be increased or decreased by appropriate stimuli.)

Axons, however, are only one element of nerve cells. Almost all axons end at synapses. A few have terminal neurosecretory endings. When a nerve impulse arrives at a synapse, it may or may not initiate another impulse in the cell with which it comes into such close proximity; only a minority of impulses reaching synapses produce corresponding impulses in another cell. Since most axons branch, especially those within the central nervous system, and since an impulse spreads equally into all branches, the number of impulses present at any given moment will obviously increase indefinitely until, and unless, they disappear somewhere. At a given synapse, the arrival of a single impulse does not usually initiate an impulse in the succeeding neuron. Synapses are those places where nerve impulses may or may not proceed from one cell to another. They are control points, places of decision, where it is determined whether a given event is to produce a particular set of consequences within the animal's body. A nerve impulse can travel in either direction along an axon, although normally it is initiated at the dendrites or cell body and thus has only one direction open. At all synapses (with a few exceptions) conduction is in one direction only, from the *presynaptic* axonal endings to either the dendrites or the cell body of the *postsynaptic* cell. (See Fig. 10-6.)

Since both axons and dendrites are highly branched in most cases, a given axon usually leads to synapses on several other cells, and any neuron in the central nervous system receives axonal endings from a number of what for it are presynaptic neurons. In a human brain or that of any other large mammal, there are more than $10^{10}$ neurons—more cells than there are people currently inhabiting the earth. Since each one has synaptic connections with many others, the number of possible connections between cells in many nervous systems is astronomical, and the task of unraveling the activity of such nerve circuits is imposing. Renewed attention is currently being paid to insect and crustacean nervous systems, especially because these have a more limited number of neurons and interconnections. Biologists hope to map out literally every event in an isolated portion of such a nervous system—a ganglion within a single abdominal segment, for example.

## *SUMMATION AND INHIBITION*

What then determines whether a given impulse crosses the synapse at which it arrives? Two important factors, *summation* and *inhibition*, are known. Their interplay is believed to account for

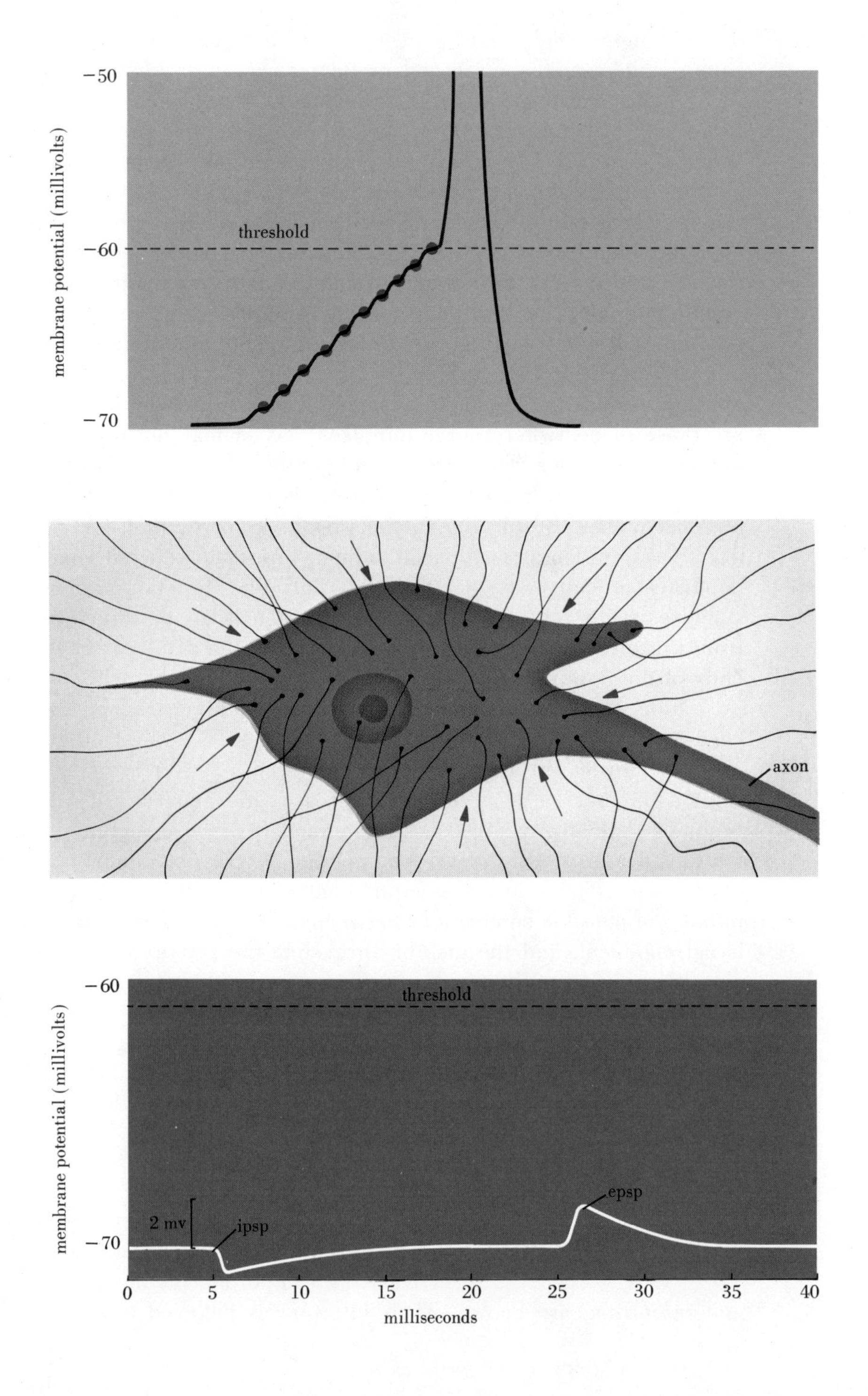
-50
-60
-70
membrane potential (millivolts)
threshold
axon
-60
-70
membrane potential (millivolts)
threshold
2 mv
ipsp
epsp
0
5
10
15
20
25
30
35
40
milliseconds

much of the functioning of nervous systems. (See Fig. 10-7.) Two or more stimulatory impulses arriving at a given synapse are more likely than one to initiate an impulse in the cell across the synaptic gap. They may arrive over the same axon within a small fraction of a second, in which case the phenomenon is called *temporal summation*, or they may reach the synapses of a given postsynaptic cell over two or more different axons, in which case the process is called *spatial summation.* A common type of temporal summation is observed when a sensory axon entering the spinal cord sends impulses at different frequencies to a synapse on a motor neuron activating a skeletal muscle. This pattern of neurons underlies many reflexes—such as the *knee jerk*, in which a sudden mechanical stretching of sense organs in the tendon of the quadriceps muscle at the knee may elicit a sharp contraction of that muscle and result in the extension of the lower leg. If the impulse frequency in the sensory neuron leading from one of the sense organs to a synapse in the spinal cord is below a critical level, no impulse is produced in the motor nerve; if it increases above this critical value, one or more impulses do travel out to the muscle. Spatial summation often occurs at the same type of synapse where nerve impulses reach it over several sensory axons within a fraction of a second. Temporal and spatial summation are essential for the transmission of impulses across synapses in the central nervous system.

Nerve impulses converging on a given synapse often produce just the opposite effect to summation—that is, one set of impulses may offset the effects of another. This phenomenon is called inhibition. At a particular synapse such as one of those involved in the knee jerk, impulses may arrive over sensory nerves at a sufficient rate to stimulate a steady series of impulses in the motor nerve. If, now, certain other neurons in the spinal cord or brain are activated and begin to send impulses over their axons to the same synapse, fewer impulses, or none at all, may occur in the motor neuron. The higher the impulse frequency in such inhibitory nerves, the less will be the transmission across the

◀ ***Fig. 10-7*** (Above) *An intracellular recording from a motor neuron to illustrate temporal summation. A few afferent neurons (synapsing with the motor neuron from which we are recording) are being stimulated repeatedly at short intervals. Each such stimulus results in a small excitatory post-synaptic potential or EPSP. These small EPSP's summate and finally bring the membrane of the motor neuron to threshold upon which an action potential is propagated.* (Center) *A cell body of a spinal motor neuron giving some idea of the number of afferent fibers which may converge on and synapse with each motor neuron.* (Below) *Inhibitory (IPSP) as well as excitatory (EPSP) post-synaptic potentials occur. A given synapse produces either one or the other but both types are usually found on a given neuron. One example appears here of each type recorded from a motor neuron cell body. (Center redrawn from "How Cells Communicate" by B. Katz, Copyright © 1961 by Scientific American, Inc. All rights reserved.)*

synapse. (In fact, the knee jerk is often inhibited. When a physician wants to distinguish between an intact but inhibited reflex as opposed to a pathologically lost reflex, he may ask the patient to undertake some trivial task such as adding a few numbers together or clasping his fingers. Such a trivial act usually cancels the inhibitory impulses and uncovers a normal reflex.) Most cell bodies of neurons or their dendrites communicate with many synaptic endings where axons of other neurons can either excite or inhibit. Hence each neuron is always subject to both excitatory and inhibitory influence. The balance between these two influences determines whether the cell in question will itself be stimulated to convey nerve impulses along its own axon.

Usually a given axonal ending is either excitatory or inhibitory; this is a property of the ending, not of the postsynaptic cell body. Terminal branches of the same axon may presumably be inhibitory at some synapses and excitatory at others.

## THE MOLECULAR BASIS OF SYNAPTIC CONDUCTION

Important information about synaptic conduction has also been gathered by measuring *synaptic potentials* (which occur in the postsynaptic cell) by means of very small electrodes placed close to a synapse while it is receiving impulses over excitatory and inhibitory axons. Synaptic potentials resemble the action potentials of an axon in some respects but are very different in others. They usually have the same polarity; that is, the arrival of excitatory impulses causes an electrode to become more positive. The synaptic potential lasts somewhat longer that the action potential—several milliseconds instead of one or a very few milliseconds. The major difference between the two is that synaptic potentials are variable in magnitude; at a given synapse they can be small or large. A low frequency of arriving impulses generates only small synaptic potentials and no action potential, in the post-synaptic cell, results. As the frequency of arriving impulses increases, the synaptic potential grows gradually until it reaches a critical value and an all-or-none impulse is suddenly generated in the postsynaptic cell. (See Fig. 10-7.) The arrival of inhibitory impulses sometimes has just the opposite effect—that is, to reduce the synaptic potential. Both excitatory and inhibitory impulses are alike in being all-or-none events, and both are represented by action potentials of the same kind. The fact that electrically identical impulses arriving over inhibitory and excitatory axons cause opposite effects at the synapse proves that something more is involved than a simple electrical process whereby the negative action potential at one cell membrane produces a similar negativity at the other and thus initiates a new nerve impulse. But we do not, as yet, know fully what *is* involved, physiologically.

The same type of interplay between excitatory and inhibitory axons occurs at the neuromuscular junctions of many muscles, including cardiac muscle of vertebrates and skeletal muscle of crustaceans, but not the skeletal muscle of vertebrates, in which there are only excitatory synapses. Although the heart in vertebrates beats spontaneously even when isolated from the rest of the animal, it is normally regulated by two sets of nerves. A chain of two sympathetic neurons, the first arising in the spinal cord and the second arising in certain of the sympathetic ganglia, carries impulses that accelerate the rate at which the heart beats. Branches of the vagus nerve (parasympathetic and also a chain of two neurons) convey impulses that cause a slowing of the heart or actually stop it altogether if the impulse frequency is sufficiently high in a sufficient number of axons.

The muscles that move the appendages of crustaceans such as lobsters receive a similar type of double innervation: one set of axons causes the muscle to contract and another tends to cause relaxation. In vertebrate skeletal muscles, inhibition occurs not at the neuromuscular junctions but at synapses in the spinal cord or in the brain. Similar principles are believed to govern all cases in which excitation and inhibition interact at synapses, but the molecular basis of the competition is most clearly established for vertebrate cardiac muscle, where it was first successfully analyzed.

The major clue to the situation was the discovery that the blood flowing out of the heart, which had just been slowed by strong stimulation of the vagus nerve, contained a substance that would exert an identical inhibitory effect on another heart into which the blood was experimentally introduced. The substance proved to be acetylcholine, which has also been found to be released in minute but effective amounts at many other types of synapses. (Acetylcholine is ordinarily destroyed rather rapidly by a special enzyme present in blood and most tissues and also present at the synaptic junction. Thus the presence of acetylcholine after stimulation of the vagus nerve is demonstrable only if the destroying enzyme, acetylcholinesterase, has been poisoned by an appropriate drug.) When the heart is caused to accelerate by similar stimulation of its sympathetic innervation, another substance, adrenalin, can be detected in the blood. Both acetylcholine and adrenalin can be added in minute quantities to cardiac muscle or to whole hearts and they produce the same effect as stimulation of the vagus nerve or sympathetic nerves, respectively. (See Fig. 10-8.)

Opposed nerve supplies are a very common type of control mechanism. Each nerve produces, at its axonal endings, very low concentrations of a highly active substance that mediates an excitatory or inhibitory effect. In many cases, however, the particular substances released at axon endings have not yet been identified. In vertebrate animals, many organs receive a dual nerve supply very similar to that

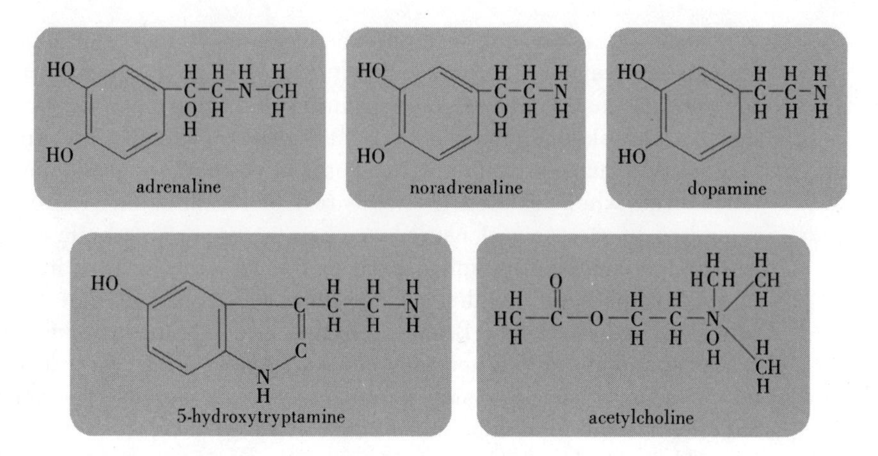


***Fig. 10-8*** *Structural formulas of several compounds which are believed to function as transmitter substances at synapses.*

which regulates the heart. The many sympathetic nerve trunks that ramify from the chain of sympathetic ganglia produce adrenalin or noradrenalin at their axonal endings. The fifth, seventh, ninth, and tenth cranial nerves—together with branches from the spinal nerves of the sacral region of the body, the parasympathetic system—usually liberate acetylcholine at their endings. These two systems, the sympathetic and the parasympathetic, together make up what is called the autonomic nervous system of vertebrate animals. (See Fig. 10-9.) Their effects can be summarized in a few significant generalizations.

The sympathetic nerves prepare an animal for vigorous action. The heartbeat accelerates; the arterioles in the digestive tract constrict but those in skeletal muscle dilate; and, although the intestinal smooth muscle relaxes, the smooth muscle of intestinal sphincters contracts, closing off one part of the digestive tract from another. These effects generally, then, shift the preparation of the animal from the quiet digestion of food to readiness for action. The bronchial smooth muscle also relaxes, and the bronchi dilate. Similar effects take place in many other organs. The parasympathetic system has opposite effects, in general, causing dilation of capillaries in the digestive tract, activating smooth muscle there while relaxing sphincter muscles and causing normal tone or constriction of the bronchi. When viewed in terms of general vegetative function as opposed to mobilization for activity, the opposition of the sympathetic and parasympathetic systems in many other parts of the body can also be understood. The two systems also often act in opposed fashion on organs that have little or nothing to do with the mobilization of energy for activity or vegetative function. In some cases, both systems function in the same direction but in different fashions, such as changing the relative concentrations of constituents of glandular secretions. There are effects of this sort on the salivary glands, the urinary bladder, and the genitalia, for example.

The antagonistic actions of sympathetic and parasympathetic systems maintain a balance throughout the whole animal; this balance is analogous to the regulated firmness of an animal's leg that results from partial contraction of opposed skeletal muscles. Neither member of the pair predominates totally except under the most unusual conditions, and their graded interplay is typical of the general strategy of control by a balance of opposed mechanisms that is widely used in all highly organized animals. Significantly, adrenalin is produced not only at the endings of sympathetic nerves but also in each of an aggregate of highly specialized sympathetic neurons, lacking axons, that make up the adrenal medulla and liberate their adrenalin directly into the blood stream. The effect of such liberated adrenalin is widespread throughout the body. The adrenalin distributed generally by the circulatory system acts on much the same organs as does the adrenalin produced by the individual sympathetic nerve endings, but there is apparently an advantage in having this alternate means of widespread and relatively massive distribution of the chemical messenger that, among other functions, alerts the animal's many tissues for vigorous activity.

Such substances as adrenalin and acetylcholine are called *neurohumors*, meaning that they are produced by nerve cells and stimu-

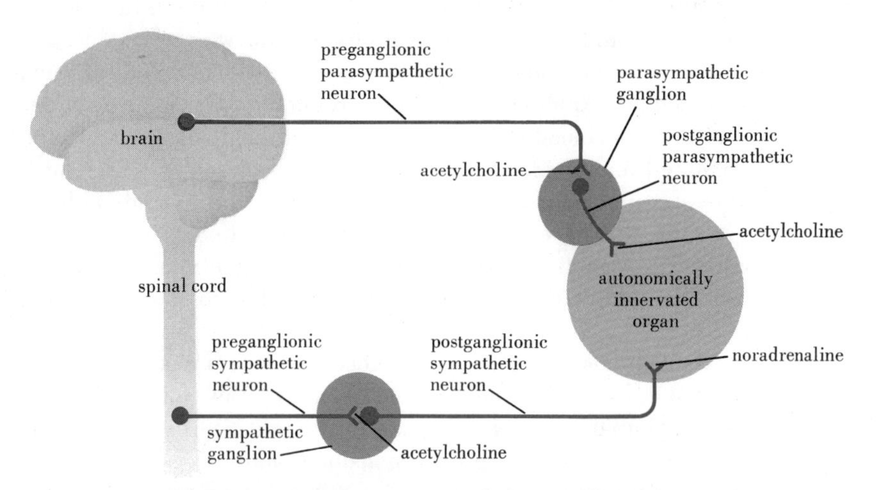


***Fig. 10-9*** *The neurochemical pattern of the vertebrate autonomic nervous system. Motor neurons to striated muscle also release acetylcholine at the motor end plate. The transmitter substances within the central nervous system remain to be fully identified.*

late or inhibit other cells. At most synapses, the *target cell* is only a few hundred angstroms distant; but, in the case of the adrenal medulla, it may be at the opposite end of the animal's body. The electron microscope has revealed hollow vesicles roughly 500 Å in diameter in the cytoplasm of presynaptic cells close to the synapse. Some of these synaptic vesicles have been shown by fragmentation and ultracentrifugation to contain acetylcholine.

## *EXPERIMENTAL TECHNIQUES*

The techniques used for elucidating the ways in which nervous systems function and how they are designed have been varied and ingenious. Basic to other studies, of course, have been systematic anatomical descriptions of the locations of groups of neurons, the destinations of their axons, and the origins of their inputs. Initially, such studies were based on gross dissections of nervous systems, the examination of sections under the microscope, the systematic staining of nerve cell bodies versus axons, and the preparation of three-dimensional models. Substantial additional information derived from the development of techniques for cutting or injuring particular tracts of axons or aggregates of nerve-cell bodies in live experimental animals and, then, identifying in their brains or spinal cords (after death) the retrograde (back to and including the cell body) as well as forward degeneration that was discovered to follow such injuries.

Many of the early insights into the localization of function derived from systematic clinical observations of men who were suffering from neurological diseases. Often the findings of detailed physical and psychological examinations were correlated with lesions (tumors, interruptions of the blood supply, or injuries) found at autopsy and accurately pinpointed anatomically. In addition, neurosurgeons, while engaged in removing tumors or repairing injuries, were occasionally able to interrupt surgery long enough to determine the effects, in unanesthetized and alert patients, of stimulation (usually electrical) of parts of the exposed brain. Since the brain itself is not pain-sensitive, such brief interruptions caused little or no discomfort to the patient.

In animals, portions of the nervous system can be cut, removed, stimulated, or inactivated (by cold or anesthetics, for example) and the resulting impairment or response evaluated. Stereotactic instruments and carefully prepared three-dimensional maps of the brain are used to place lesions precisely and often in very elegant ways (with reversible effects, for example). Electrodes can also be implanted in regions of the central nervous system or into individual neurons and often into or in the vincinity of sense organs to record the electrical activity that occurs. Usually one observes such electrical activity in response to some controlled event to which the animal is exposed. For example, a variety of wavelengths of light may be directed at selected portions of the retina or sounds of given structure may be fed into one or both ears simultaneously or sequentially. Electrodes in visual or acoustic interpretation centers will then record electrical activity which can (given sufficient patience and ingenuity) be associated with the physical events that have impinged on the animal.

Substantially, these techniques concern themselves with increasingly detailed descriptions of the anatomical interconnections of the neurons and with the display of activity of the neurons in response to or in producing (in the case of effector neurons) describable events.

Another group of experimental techniques deals with the input and output of the nervous system and uses inductive reasoning to construe the internal events. For example, the details of learning in animals have been studied using operant conditioning. Here, essentially, an animal is rewarded for performing a given task (or punished for not doing so) and all of the details surrounding its ability to conform can be studied, piece by piece, by experimental variations. Pigeons have been taught to play a form of ping-pong this way—and although this may seem at first glance a triviality, it actually is not. Some of the higher functions of nervous systems such as learning, memory, and judgment may be analyzed this way.

Operant conditioning and related techniques may be used, in addition, to measure perception (the psychological appreciation of some physical event such as the presence of light or the playing of the musical tone middle C) in animals. Not only can their visual or acoustic capacity be assessed in this way, but also their ability to discriminate fine differences. Thus, some of the input of nervous systems can be quantified. In fact, many of the best data on sensory perception seem to be from such experiments.

Finally, the natural or laboratory behavior patterns of men and animals can be described and analyzed. For example, the phenomenon of territoriality can be explored extensively by observations on its occurrence in nature, by attempts to induce or alter territorial behavior in the laboratory under controlled conditions, and by observations of many species, identifying the roles served and the evolution of the phenomenon from the comparative data.

Initial studies of nervous systems almost entirely involved vertebrates, especially mammals. The suitability and utility of invertebrates was only slowly recognized. Not only are their nervous systems small, involving fewer cells (often literally countable and individually recognizable) but many are so organized as to permit the isolation of small sections, such as a segmental ganglion. Furthermore, invertebrates are often abundantly available, inexpensive to keep in large numbers, and easy to work with. Decapitated insects, for example, may survive for two years if properly cared for. Perhaps, the greatest contribution that the invertebrates have made, however, has been in their experimental utility in elucidating the nature and role of neurosecretion (to be discussed in the next section). Here an early hero was the roach; later ones have included silkworms and an exotic bedbug. These and many more have contributed substantially to the advances in our understanding of vertebrate neurosecretion as well as the ways in which nervous systems, in general, work. Neurobiology—using some of the methods described, as well as some yet to be designed—promises to be the explosively productive field of the next few decades.

## *ENDOCRINE CONTROL SYSTEMS*

Close associations between systems of *nervous* and *hormonal control* are being demonstrated in more and more cases. This fact leads us naturally to a consideration of *endocrine glands* that produce hormones as their principle function. The *medulla* of the *adrenal gland* is one of a number of endocrine glands, so named because they discharge their products into the bloodstream rather than into ducts leading to the outside surface or into the digestive or reproductive tracts or their derivatives. The function of the hormones thus secreted is to affect, in some way, one or more target organs or tissues remote from the endocrine gland concerned. Acetylcholine is considered a neurohumor but not a hormone, whereas noradrenalin is both. In all cases, a given hormone is present in the blood in only very minute quantities, which may require special techniques for detection. Detection by *bioassay* is the most common. Here, a living organism or part of a living organism, usually of a standardized strain, known to be very responsive in some measurable way to a given hormone, is often injected with a sample of blood, serum, urine, or tissue extract (or fed such a sample or exposed to the sample in some equivalent manner) and is, then, observed for the appropriate reaction. In studying insect growth and metamorphosis, for example, isolated (by ligation) blowfly abdomens (which survive for long periods if cared for properly but never metamorphose because they lack the hormones required for metamorphosis) may be injected with material believed to contain the molting hormone. If this hormone, *ecdysone*, is present, the ligated abdomen metamorphoses into a pupa. The quantity required to cause puparium formation in 70 percent of blowfly larvae, genus *Calliphora*, is called one *Calliphora* unit. Such quantities are usually expressed as units of activity since the actual quantity of hormone present cannot be expressed in grams or similar units without being extracted and chemically identified.

A hormone is defined by the following criteria: (1) it is produced in a particular endocrine gland; (2) removal of the gland causes the failure of some other organ or organs to develop or function normally (the normal effect may be stimulatory or inhibitory); and (3) in the absence of the endocrine gland, artificial replacement of the hormone in the body restores the target organ to normal size and functional capacity. Several hormones have very general effects throughout the body. These include several from the *cortex* of the *adrenal glands* that regulate ionic balance and carbohydrate and protein metabolism, and in many other ways, not yet clearly understood, promote the general vigor of the animal. They are also involved in wound healing and immune reactions. The *thyroid gland* produces *thyroxin,* which increases the general metabolic rate of the animal and is also essential for metamorphosis in amphibian tadpoles. Endocrine cells located in the *pancreas* (a gland also containing other cells that secrete digestive enzymes into the intestine) produce *insulin,* which is necessary for the utilization of glucose by cells throughout the body. The hormone of the *parathyroid glands* regulates calcium and phosphorus metabolism. Little is yet known of how these hormone molecules produce their observed effects. In some few cases, the influence is known to be on enzyme systems.

Another group of hormones is produced by the *anterior pituitary gland,* which lies adjacent to the hypothalamus of the brain. The production of these hormones appears to be governed by *neurosecretions* of the hypothalamus released into the *hypophyseal portal system* by hypothalamic axons ending near the portal system capillaries. (See Fig. 10-10.) Probably six independent regulatory neurosecretions are produced, one to govern the production of each of the anterior pituitary hormones. One appears to be inhibitory in function and the other five stimulatory. The anterior pituitary produces at least six distinct hormones, all of which appear to be proteins. These are: *growth hormone* (called *SH* because of its other name, somatotropic hormone), which regulates the general growth of young animals, especially of the long bones; *adrenocorticotropic hormone (ACTH),* which regulates the production of some of the adrenal cortical hormones; *thyrotropic hormone (TSH),* which regulates the activity of the thyroid gland; *follicle-stimulating hormone (FSH),* called *interstitial cell-stimulating hormone (ICSH)* in males, which regulates the growth and maturation of the follicles and ova in the ovaries or of the interstitial cells of the testes; *luteinizing hormone (LH),* which regulates the activity of the corpus luteum of the ovary in females and influences spermatogenesis in males; and *lactogenic hormone* or *luteotropic hormone (LTH),* which regulates the production of milk by the mammary glands in female mammals. All of these hormones, where studied, have proved to have similar and often additional duties in other vertebrates.

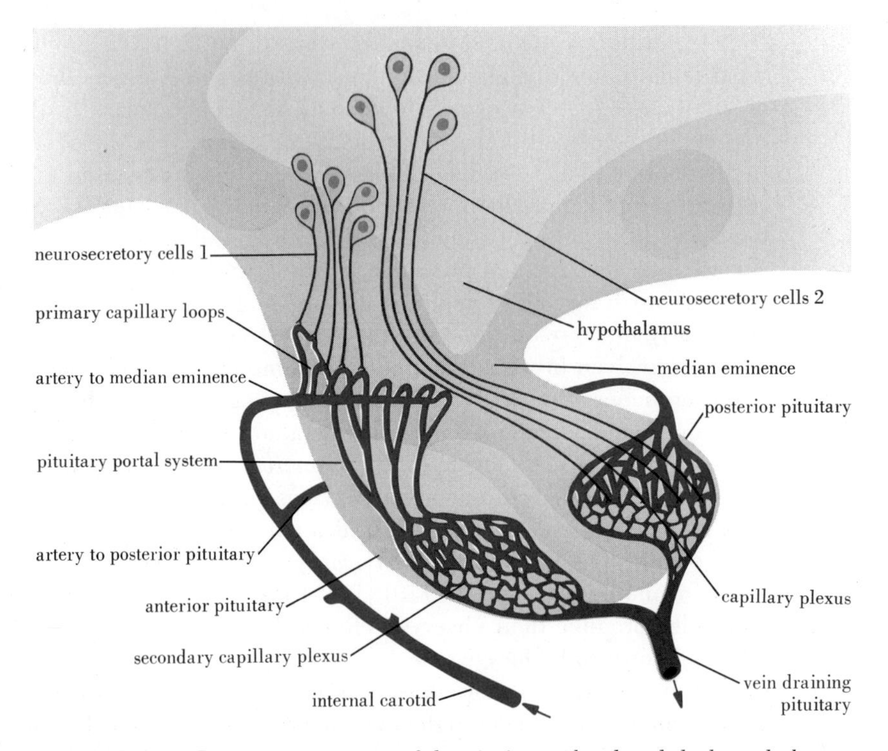


***Fig. 10-10*** *Diagrammatic view of the pituitary gland and the hypothalamus of a rabbit with the origin and distribution of the hypothalamic neurosecretory cells and the circulation simplified. The axons of neurosecretory cells in the hypothalamus extend either to the median eminence (where they end in association with capillary loops of the pituitary portal system) or into the posterior pituitary (where they end close to the "normal" capillaries present there). The primary capillaries of the median eminence coalesce (carrying blood charged with the neurosecretory products of the neurosecretory cells ending there) to form larger vessels which flow into the anterior pituitary and there divide to form a secondary plexus of capillaries. Thus, the only arterial blood supply to the anterior pituitary has already passed through a capillary system in the median eminence. This pituitary portal system serves as a private delivery service for neurosecretory substances of the median eminence to the arterior pituitary. The venous drainage of both lobes of the pituitary is normal. (Adapted from Jenkins, P. M.* Animal Hormones, A Comparative Survey, *Pergamon Press, New York, 1962.)*

The *gonads,* themselves, are also endocrine glands. The *ovaries* in mammals produce, cyclically, *estrogens* (associated with follicular growth) and *progesterones* (associated with the *corpus luteum,* which develops from a follicle after the ovum has been released). Estrogens and progesterones act together to prepare the uterus for reception of a zygote and to maintain the uterus in condition to house and nourish

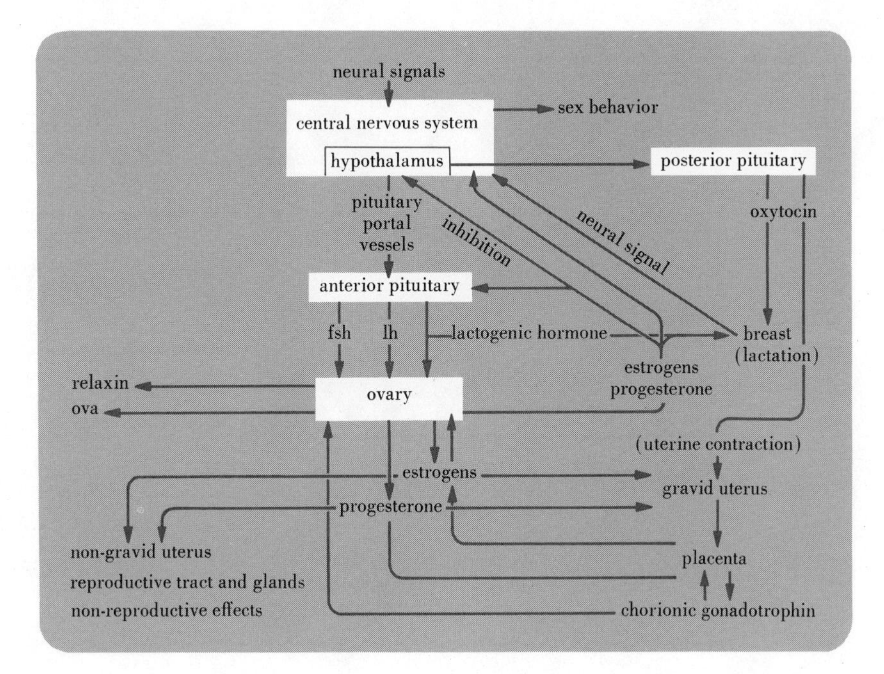


***Fig. 10-11*** *An outline of the neuroendocrine control of reproduction in a female mammal.*

the embryo. (See Fig. 10-11.) These hormones also regulate many other features of the body—such as hair and fat distribution, the condition of the external genitalia, sexual receptiveness, and so on—that can be grouped as reproductive functions or secondary sex characters. Contraceptive pills often consist of these hormones, or synthetic substitutes, which alter the cyclical events sufficiently to prevent ovulation or to interfere with the preparation of the uterus for implantation.

The *interstitial cells* of the *testes* produce *testosterone*, which is responsible for male secondary sex characters such as hair distribution, sexual odors, the activity of the prostate gland, and male behavior in courtship and mating including interest in and competence in copulating successfully. The significance of secondary sex characters such as hair distribution is by no means entirely clear. In some cases these characters announce the sex of the owner to other members of the same species. In yet other cases, these characters may serve as signals for appropriate episodes of courtship or mating behavior. Other secondary sex characters function in the mechanics of mating. (See Fig. 10-12.)

The *placenta* is an important endocrine organ which varies in its activity from species to species but in some can be shown to produce *gonadotropins*, such as TSH, LH, estrogens, progesterones, and lacto-

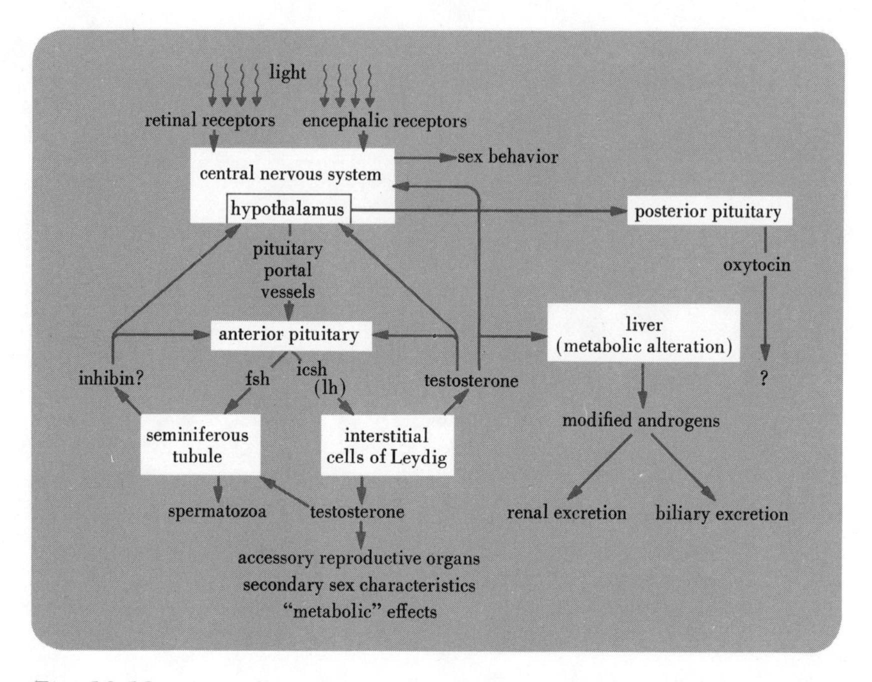


***Fig. 10-12*** *An outline of the neuroendocrine control of reproduction in a male mammal. The input shown, light, or photoperiod is meant to serve as an example. Other sensory inputs may also trigger or influence the interactions shown.*

genic hormone. In many mammals, the placenta is eaten by the mother at the time of delivery and is believed to influence, by its hormone content, the contractions of the uterus as well as the subsequent production of milk.

This does not complete the list of vertebrate hormones any more than other organ systems have been described completely in this brief book. The relationship between the anterior pituitary and the ovaries will be briefly considered because it illustrates an important general principle of endocrine control. Not only do the gonadotropins of the anterior pituitary stimulate the ovarian cells to produce estrogens and progesterones which affect the secondary sex characters and the uterus, in particular, but the estrogens and progesterones produced by the ovarian cells in turn affect the activity of the anterior pituitary. (See Fig. 10-13.) As far as we know at present, the level of circulating estrogens and progesterones is monitored by some mechanism that seems to lie in the hypothalamus of the brain. When the estrogen and progesterone levels, respectively, rise, the hypothalamic neurosecretory products affecting the anterior pituitary production of FSH and LH, respectively, are inhibited. Thus with the rise in level of circulating

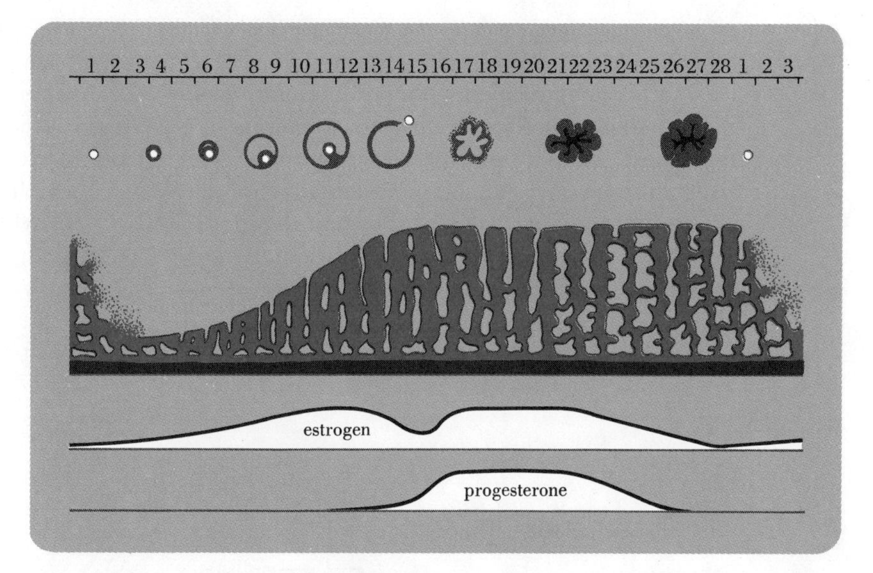


***Fig. 10-13*** *A diagram correlating ovarian and uterine events with the levels of circulating ovarian hormones during a 28 day menstrual cycle in a woman. Slightly more than one cycle is shown. Day 1 conventionally is the first day of menstrual bleeding. Ovarian changes are shown above and progress from follicular growth through ovulation and the subsequent formation and eventual regression of the corpus luteum. The endometrium of the uterus shows a progressive increase in thickness, vascularity, and the complexity of glands. The rise in estrogen level is associated with follicular development. Ovulation occurs at a time of falling estrogen and rising progesterone. High estrogen and progesterone levels support the complex structure of the uterine endometrium but as the levels of these hormones drop (if pregnancy does not occur), the uterine endometrium is sloughed off and the cycle starts anew. If pregnancy occurs in a given cycle, the fertilized egg, after a few days of free development, implants in the richly prepared endometrium.*

estrogen, the stimulating effect on the anterior pituitary by the hypothalamus to produce follicle-stimulating hormone is discontinued. The level of follicle-stimulating hormone then drops and, as a result, the rate of estrogen production also drops. The sequence of events appears to be similar for progesterones and luteinizing hormone. Thus, there is a double cycle known to involve the hypothalamus, the anterior pituitary, and the ovaries, each of these organs producing at least two products of hormonal character. This cyclical activity is responsible for the cyclical nature of estrus in mammals and for the somewhat specialized estrus cycle, called the menstrual cycle, in man.

The hormone, *oxytocin*, is also secreted by hypothalamic axons, but directly into the systemic blood, not into the hypophyseal portal system. This is another example of neurosecretion. Oxytocin causes uterine contractions and *let-down* or *ejection* of milk. (See Fig. 10-14.) In many mammals the physical stimulation of copulation causes the release of oxytocin, which is stored in quantity near the axonal endings of hypothalamic neurons. These axon endings are gathered in a separate cluster, called the *posterior pituitary*. The release of oxytocin causes uterine and oviduct contractions, in the female, which may facilitate the prompt arrival of sperm at the interior end of the oviducts where the ova usually await fertilization. At the same time that the stimulation of copulation causes the release of oxytocin, in many mammals such as cats and rabbits it also stimulates the release of a hypothalamic factor, which acts through the hypophyseal portal system on

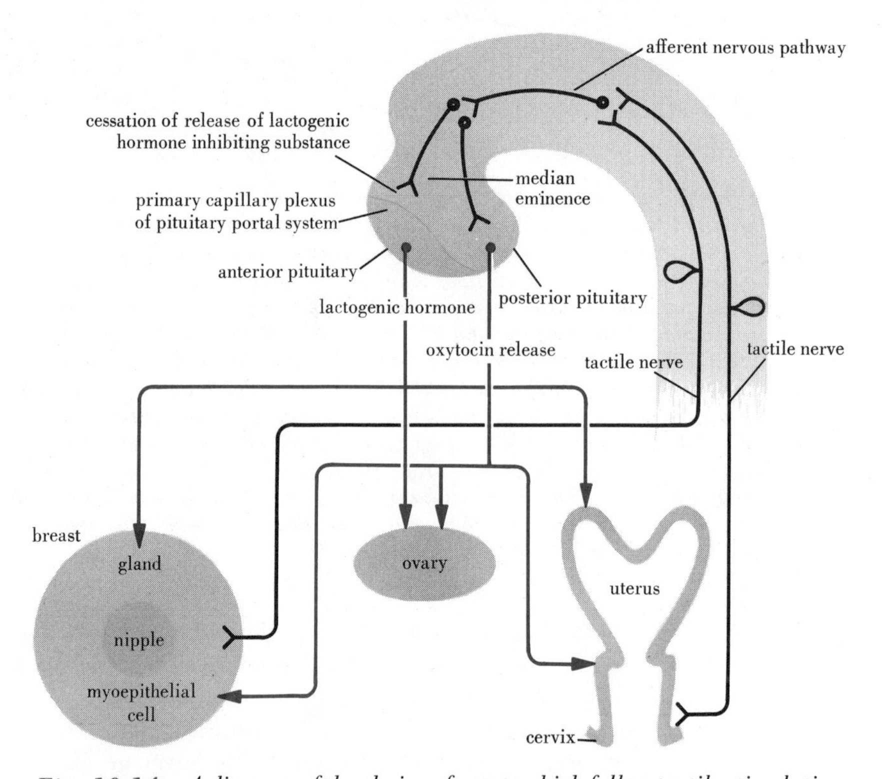


***Fig. 10-14*** *A diagram of the chains of events which follow tactile stimulation of the nipple of the mammary gland of a lactating mammal (as by the sucking movements of the baby, or of the cervix of the uterus in some estrous mammals (during copulation in cats or in rabbits, for example). The proximal portion of the chain is neural, reaching as far as the neurosecretory cells of the hypothalamus. The next step is neurosecretory. In both cases, oxytoxin may be released and acts directly on several target organs. In addition, the release of lactogenic hormone inhibiting substance is stopped. This permits production of lactogenic hormone by the anterior pituitary. Thus a third portion of this branch of the chain of events is endocrine, again with a number of ultimate target organs.*

the anterior pituitary. The anterior pituitary production of FSH and LH then signals ovulation. In such mammals ovulation occurs only following the physical stimulation of copulation, which is reported to the hypothalamus via sensory nerves. Oxytocin is also released in males at the time of copulation, but its function is not yet well understood. Oxytocin is also released in female mammals as a reflex response to the sucking of the young on the mammary nipples. The sensation of sucking is relayed to the brain by sensory nerves. The hypothalamus responds by releasing oxytocin, which causes certain cells of the mammary glands (the *myoepithelial* cells) to contract, forcing milk out of the glandular canals into the major ducts of the nipple from which the milk then flows freely. These myoepithelial cells are contractile musclelike cells surrounding the milk-secreting cells of the mammary glands. Lacking the contraction of the myoepithelial cells, milk will not flow. This reflex can be easily conditioned so that the let-down of milk may occur with experience in response to the sight of the young, the sound the baby makes, or even to soft music or colored lights. Such conditioning is exploited commercially to facilitate the milking of cows. The *production* of milk as opposed to its let-down is controlled by lactogenic hormone. The same sensory input which causes a release of oxytocin, the sucking on the nipple by the baby, also inhibits the production or release of the lactogenic-hormone-inhibiting substance of the hypothalamus. Thus the anterior pituitary is released from inhibition in this respect and increases its production of lactogenic hormone, which thereby maintains the production of milk by the mammary glands. When the mother ceases to suckle the young, at the time of weaning, this hormonal sequence is broken and lactogenic hormone production is inhibited. Here we have two excellent examples of the interactions of nervous transmission, neurosecretion, and endocrine function.

Nerve impulses serve for the rapid conveying of specific signals to particular cells, including both positive excitation and negative inhibition. To express equally succinctly the roles of neurosecretions is still a task for the future. They can certainly by described, but it presents a problem currently to distinguish rigorously their roles from those of the nervous system. Adrenalin acts highly specifically locally and also quite generally when systemic. Oxytocin and antidiuretic hormone have very clear target organs at a distance. They seem to be capable of quantitative action when integrated into a feedback system. The hypothalamic factors governing anterior pituitary function are delivered locally and quantitatively and their effects (with feedback arrangements) again seem to be quantitative. The analysis of these control and integration systems (including, of course, other endocrine organs) promises to be a most challenging and productive field for research for some time to come.

## THE ORIGIN OF NERVOUS EXCITATION

At this point it is appropriate to inquire where all the activity of the nervous system originates. Many nerve impulses derive from sense organs and flow into the central nervous system along sensory nerves. The major types of receptor cells are important because they are the only known channels of information between the outside world and the central nervous system. Many receptors are concentrated into elaborate sense organs such as the eye or ear, but roughly as many others are distributed widely throughout many tissues of the animal's body. The simplest are undifferentiated or relatively undifferentiated nerve endings that are common in the skin and sensitive to mechanical deformation, temperature changes, and relatively pronounced changes in their chemical environment. The more vigorous such stimuli are, the higher the frequency of nerve impulses that travel along these neurons to the brain or spinal cord. In ourselves and presumably in other animals, many of these impulses produce sensations of pain when sufficient numbers of neurons carry impulses at high frequencies. Increased sensitivity to minute amounts of energy of specific types is attained by specialized receptor cells. When such receptor cells are organized into sense organs, the resulting sensitivity sometimes approaches limits set by the physical nature of the stimulus energy. For example, the rod cells of vertebrate eyes may respond to a single photon.

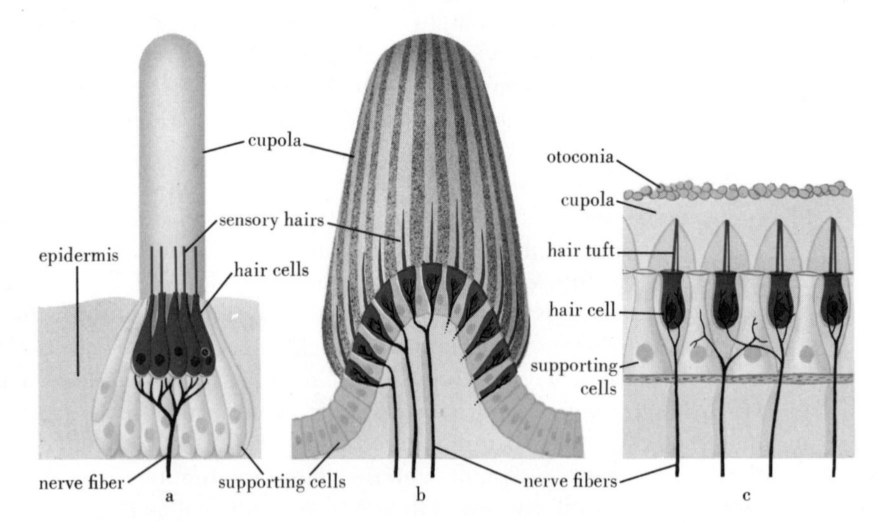


***Fig. 10-15*** *Diagrams of some receptors of the acoustic-lateralis system.* (a) *A superficial neuromast of a bony fish (after Dijkgraaf, 1963).* (b) *A mammalian crista and* (c) *a mammalian macula, both parts of the inner ear associated with the measurement of acceleration and gravity (after Netter, 1953).*

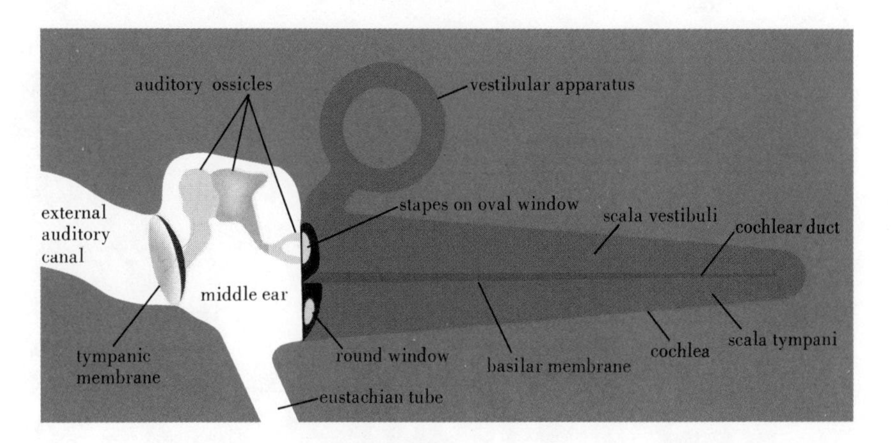


***Fig. 10-16*** *A diagram of the anatomy of the acoustic portion of a mammal's ear. The spirally coiled cochlea has been unrolled to clarify the relationships between the two fluid filled compartments—the cochlear duct, which encloses the acoustic sense organ, and the continuous scalae vestibuli and tympani, which connect with the middle ear and conduct sound waves past the sense organ.*

Many fishes have sensitive chemical receptors lining an olfactory pit, located on the head, through which water circulates as the fish swims about. In some cases the fish recognizes a particular odor at such a low concentration that only one or a very few molecules could have reached any one receptor cell. The most widely accepted current hypothesis to account for the homing ability of such migratory fish as salmon states that the fish recognize, by olfaction, the particular water draining from their birthplace.

In the ears and related sense organs of vertebrate animals, very small oscillatory pressure changes that constitute sound waves in the surrounding air or water are conducted to specialized *hair cells*, from which cilia project. The cilia, which do not move spontaneously, are usually embedded in a mass of gelatinous material, the *cupola*, that floats in a fluid surrounding the hair cells. (See Fig. 10-15.) As a result, sound waves reaching the vicinity of the hair cell move the gelatinous cupola and, in so doing, deform the cell surface ever so slightly where the cilium emerges. These cells are extraordinarily sensitive, approaching the theoretical limit of detecting random molecular movement. The sensitivity of ears could not be much increased without the animal's beginning to hear the random thermal agitation of the molecules comprising its ear. (See Figs. 10-16 and 10-17.)

One significant example of the manner in which auditory signals are important to certain animals is the biological sonar systems used by bats while flying in darkness. Orientation sounds, lasting only a millisecond or so, and containing frequencies that are ordinarily above the

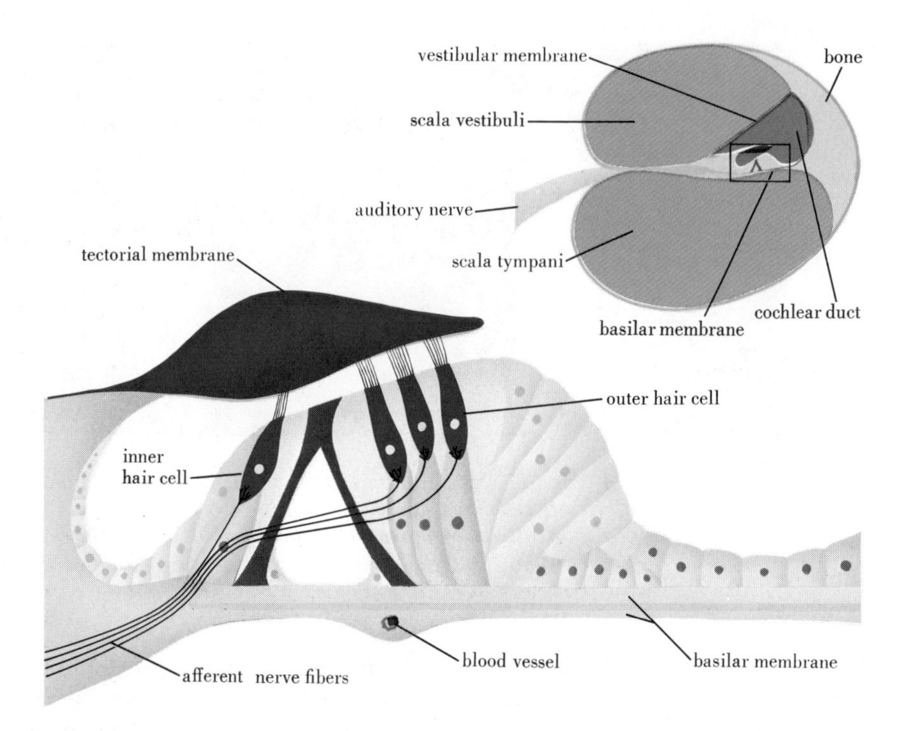


***Fig. 10-17*** *Diagrams at two different levels of magnification of a cross section of the cochlea.* (Below) *The aggregation of the hair cells into the groups, their innervation, and the relationship of the hair cells and their cilia to the tectorial membrane (or cupola) and to the basilar membrane.*

range of human hearing, are emitted to probe the animal's environment. Echoes from obstacles or from insect prey activate the two ears in much the way that a man hears sounds of lower frequencies. The specialized auditory portions of the brain (which in the medulla and midbrain of bats bulk larger than all other sensory systems) analyze these sonar echoes within a fraction of a second; thus the bat can dodge a twig or capture a moth. A total pursuit, in fact, from detection of an insect to its ingestion may take less than 700 msec. Whales and dolphins employ a similar form of biological sonar based upon sound waves under water. This type of *active acoustic orientation*, known as *echolocation*, is especially impressive not only because of the faint signals to which bats and whales react, but because crucial behavioral decisions are reached within very short periods of time to orient the animal's locomotion and, in particular, its pursuit of elusive, moving prey.

A remarkable example of how a shift in emphasis among sense organs interacts with other aspects of an animal's life is presented by the bird genus *Collocalia*, the swiftlets of Southeast Asia. The less specialized birds of this genus orient visually and build nests consisting chiefly of vegetation, pasted together with salivary secretions, on vertical rocky surfaces such as cliff faces. The vegetation to be used for nest building is carried in the beak. At least three species of these swiftlets have evolved a sonar or echolocation system similar to that of bats which they use, however, not for hunting or for general navigation but for flying in the dark caves to which they have shifted their nesting, apparently for the advantages gained versus predators and bad weather. They produce their sonar signals through their mouths, however, and this interferes with their carrying nesting material in their beaks. One species continues to build an ancestrally conservative nest but carries the building materials with its feet. Another species carries in no nesting material at all but builds its nest of its own feathers. The third species, the one whose nests are exploited commercially for bird's-nest soup, builds its nests entirely of secretions produced by a buccal gland. Thus these species have eliminated the conflict between the use of the mouth for producing sonar signals and for carrying building materials at the same time.

Eyes are no less remarkable than ears in their sensitivity. Transparent tissues that grow to have the exact shape required of image-forming lens systems focus an image of the outside world on a retina or mosaic of visual receptor cells. In parts of the eyes of many vertebrates, these visual receptors are packed within about 3 $\mu$ of each other so that they can be individually stimulated by fine details of the focused image down to the limits of resolution set by the wave lengths of visible light. There are two basic types of visual receptors in the retinas of vertebrates—rod and cones. Rods are almost universally present (they appear to be lacking in many diurnal birds) and are designed for high sensitivity under low-light conditions. They not only have a lower threshold of sensitivity than cones but are usually so linked to the central nervous system that the responses of up to several hundreds or thousands summate. Cones are designed for precise vision under good lighting conditions. Their signals are frequently reported to the central nervous system as individuals or as small groups. Cones are also responsible for color vision. Color (which depends on the wavelength of the received light) is not distinguished by rods. Rods and cones are both modified *ependymal cells* of the central nervous system. Embryologically speaking, the retina is derived as an outpocketing from the brain. The visual cells themselves are derived from a single layer of cells that normally line the central canal of the central nervous system. Such cells are ciliated—as, indeed, are the rods and cones. The highly modified cilia of rods and cones are called *outer segments*. The outer segments of rods are packed with organized layers of a photosensitive pigment, *rhodopsin*. (See Fig. 10-18.) A closely related molecule, *porphyropsin*,

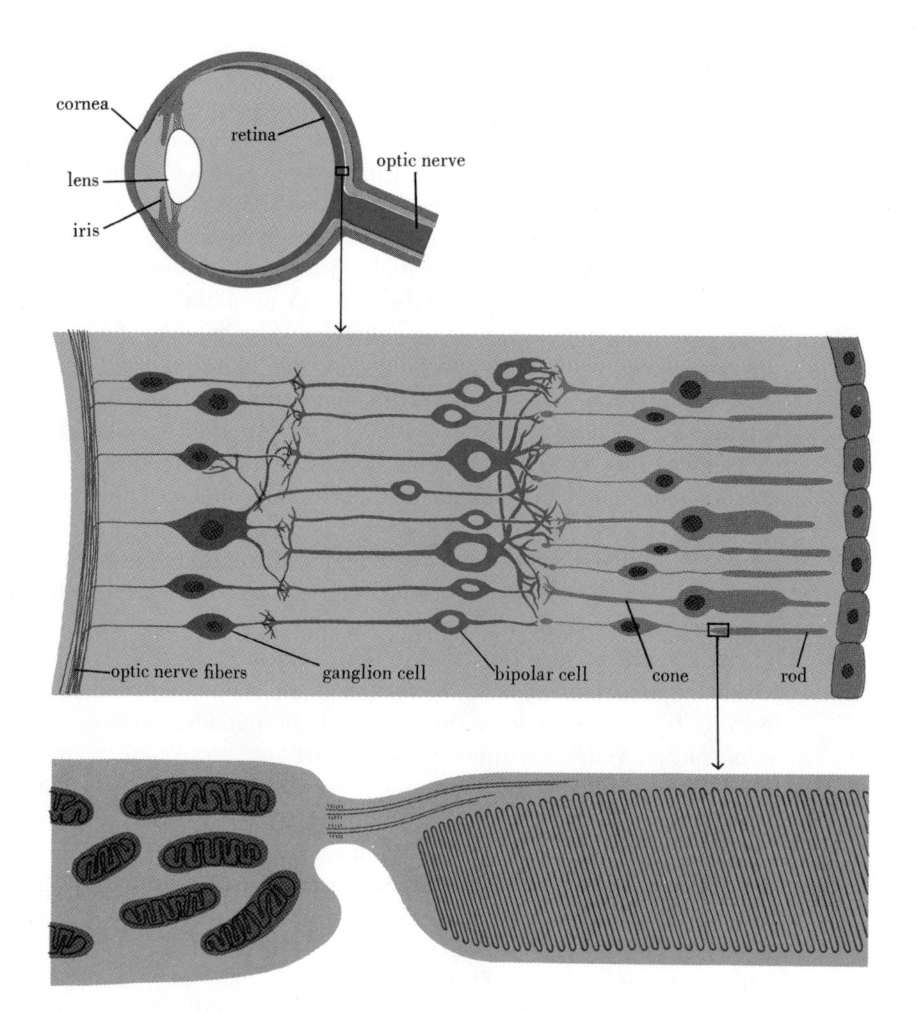


***Fig. 10-18*** *Three views of the eye and its components in a mammal.* (Above) *A sagital section of the eye ball and the optic nerve.* (Middle) *A diagram of the cell layers which make up the retina. The receptors, the rods and cones, are in the interior of the retina and face away from the entering light. A receptor, especially a cone in the central fovea, may have a "private" bipolar cell. Generally, a few cones or many rods converge on each bipolar cell and many bipolar cells converge on each ganglion cell. Many bipolar cells connect to more than one ganglion cell so that they enter into different convergences. In addition, certain other cells in the retina make lateral connections (one is shown here in the bipolar layer).* (Below) *A diagram of part of a rod taken from electron micographs. The visual pigment, rhodopsin, is arranged in stacked membranes (like a pile of phonograph records) in the outer segment. A modified cilium can be seen in the stalk connecting the rod outer segment to the cell body proper. The cilium may have some function in conducting the signal of excitation from the outer segment to the cell body. (Adapted from W. Telfer and D. Kennedy,* The Biology of Organisms, *John Wiley & Sons, Inc., 1965.)*

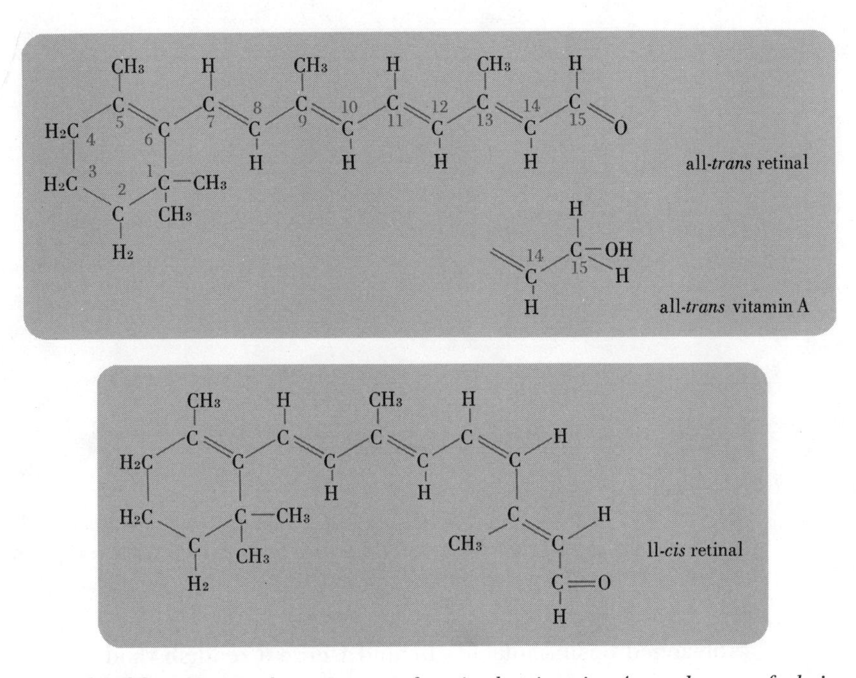


***Fig. 10-19*** *Chemical structure of retinal, vitamin A, and two of their isomers. Vitamin A is the alcohol of the aldehyde, retinal. One of the breakdown products, in vertebrates, of rhodopsin bleached by light is all-*trans *retinal. In order to resynthesize rhodopsin, however, the retinal must be isomerized to the 11-*cis *form. Retinal$_2$ and vitamin A$_2$, differing by the presence of an additional double bond in the ring, occur in some fish and amphibians and make up part of a parallel porpshyropsin system.*

occurs in the rods of some fish and amphibians. Rhodopsin molecules consist of a large protein portion (called an *opsin*) attached to a small carotenoid molecule, which is called *retinal* when it exists as a separate unit. The retinal molecule, which is *vitamin A aldehyde*, can exist in any of five different *cis-trans* isomers. (See Fig. 10-19.) These isomers differ in stereo configuration; that is, their actual three-dimensional shapes are different. Apparently only one particular shape fits, in nature, on to the active part of the opsin molecule. When this particular isomer (11-*cis*) is mixed with opsin in a test tube, the reaction to combine to make rhodopsin proceeds spontaneously to completion (it is an energy-yielding reaction). (See Fig. 10-20.) But when a quantum of light, of sufficient energy, is absorbed by this molecule, the 11-*cis* retinal segment is isomerized (deriving the energy of activation from the quantum of light) to the all-*trans* configuration. This shape of retinal no longer fits onto the active site of the opsin molecule. (See Fig. 10-21.) The retinal portion eventually snaps out of position and separates, leaving

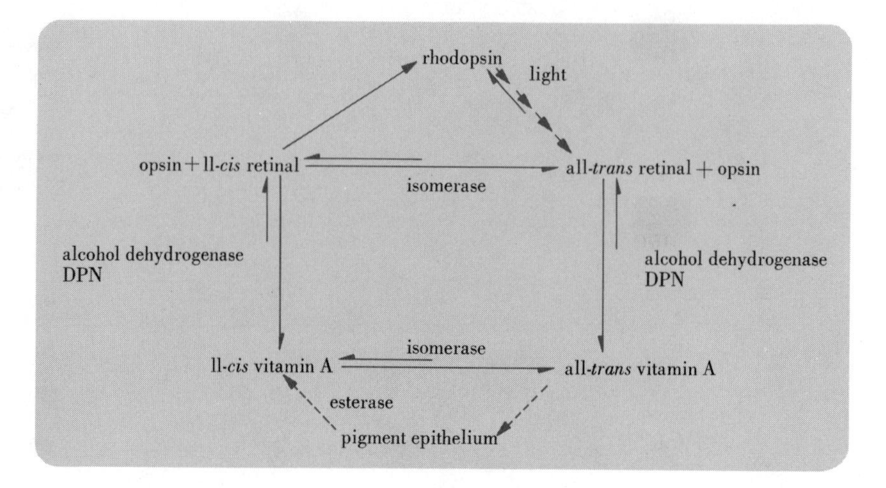


***Fig. 10-20*** *Pathways of bleaching and reconstituting rhodopsin. The intermediate steps in the bleaching reaction have been summarized and represented only by a series of short arrows.*

a bleached opsin molecule behind where a reddish rhodopsin had previously been. The opsin by itself is not light-sensitive. In order to respond to a subsequent light signal, any of millions of other rhodopsin molecules may act, but in time all of these would be split and insensitive. Rhodopsin must be resynthesized. Such resynthesis requires the enzymic conversion of all-*trans* retinal into the 11-*cis* form. An enzyme present in the retina, *retinal isomerase*, is believed to serve this function. Retinal is in equilibrium with *vitamin A*. Vitamin A molecules—which must be ingested and cannot, in general, be synthesized by animals

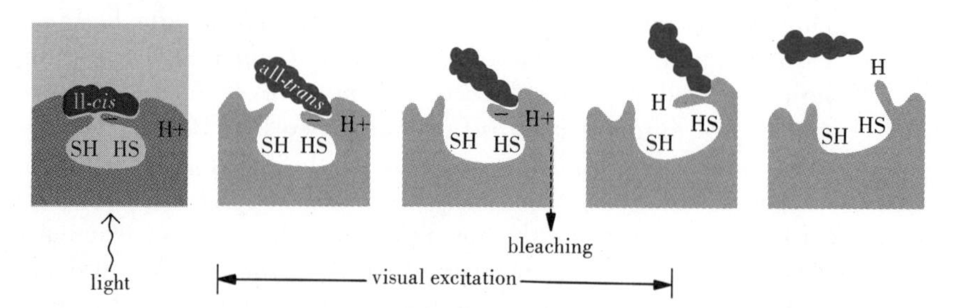


***Fig. 10-21*** *A hypothetic view of some of the events in the bleaching of rhodopsin. The only action of light is to isomerize 11-*cis *to all-*trans *retinal. Thereafter thermal reactions cause configurational changes in the opsin and eventually retinal is hydrolyzed and detaches from the opsin. Visual excitation occurs during the period shown but the mechanism is, as yet, not known. (Adapted from Wald, G. and P. K. Brown,* Cold Spring Harbor Symp. Quant. Biol., *30, 1965.)*

(hence called a vitamin)—are stored in fat, in the liver, and in the pigment epithelium of the eye. All of these sites are in equilibrium, via the blood, with the vitamin A in the retina. If there is a lack of vitamin A, vision deteriorates—night or rod vision apparently before cone vision. Such a degree of vitamin A deficiency, however, is hard to achieve, even experimentally, because of the ubiquity of vitamin A and its precursors. The change in shape of the retinal portion of the rhodopsin molecule may be the event which initiates an excitatory response in the receptor. No one knows how these events are physicochemically linked to the initiation of the nerve impulse. The rhodopsins do appear, however, to be arranged in membrane layers that are only two molecules thick. Perhaps, the isomerization of the retinal portion leaves a gap or hole in the membrane, changing its permeability to ions or altering it in some other physicochemical sense. Such a change in permeability could conceivably result in the initiation of a nerve impulse.

The maximum sensitivity of rhodopsin (the wavelength at which it absorbs the most light) is in the blue-green part of the visible spectrum. (See Fig. 10-22.) Porphyropsin's absorption maximum is shifted some 20 mμ (in wavelength) toward the red. The cones of animals that utilize rhodopsin in their rods contain a pigment called *iodopsin* consisting of retinal combined with a somewhat different protein molecule—a *cone opsin.* (See Fig. 10-23.) The porphyropsin group of vertebrates use a related cone pigment, *cyanopsin.* The molecules of cone pigment

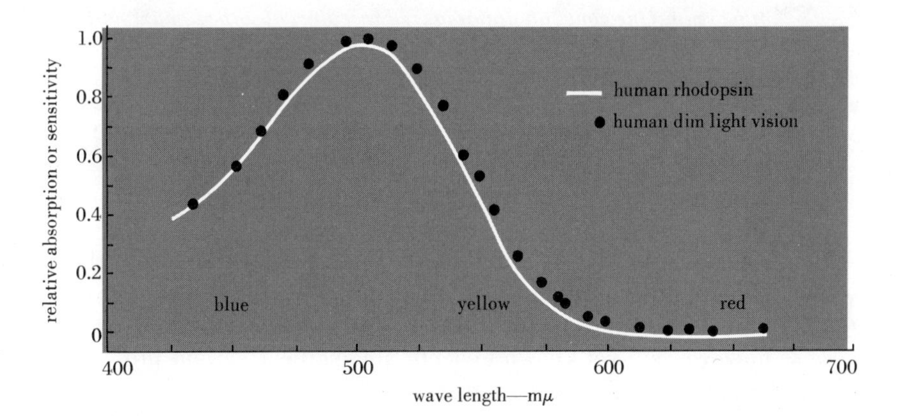


***Fig. 10-22*** *The absorption of a rhodopsin solution compared with the visual threshold of a man at different wavelengths of (dim) light. Using dim light eliminates the cones which have a much higher threshold. The coincidence of the two curves strongly supports the hypothesis that rhodopsin is the visual pigment for dim light in man. That is, the absorption spectrum of the pigment is almost identical (within experimental error) with the threshold curve. (From G. Wald and P. Brown,* Science, *127, 1958.)*

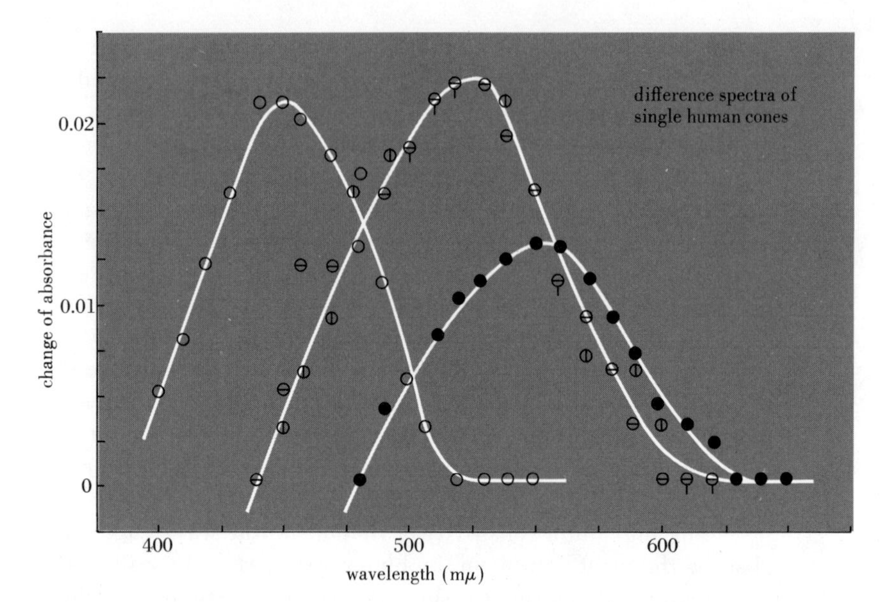


***Fig. 10-23*** *Difference spectra of the visual pigments of several individual parafoveal cones in man. (The data points for each cone are shown differently.) In each case, the absorption spectrum was measured first under dark adapted conditions and then after bleaching the pigment with a brilliant flash of yellow light. The curves shown are the differences between these two measured spectra and are thought to approximate the absorption spectra of the original unbleached pigments. One of these cones, apparently a blue receptor, absorbed maximally at 450 mμ. Two cones, apparently green receptors, absorb maximally at 525 mμ. One cone, an apparent red receptor, absorbed maximally at 555 mμ. (After Brown and Wald, 1964.)*

are apparently far less densely packed than are the rod-pigment molecules. The opsin portion of all of these pigments varies distinctly from species to species and from order to order but the retinal fragment occurs in only two variants, *retinal*$_1$ and *retinal*$_2$. Each particular rhodopsin is slightly different because of the variations in the opsin as, indeed, other proteins such as insulin, growth hormone, or hemoglobin differ from species to species. No obvious variations in functions, however, are yet known to occur among these various protein molecules in closely related vertebrates.

Retinal-based rhodopsin has also now been identified as the visual pigment in the eyes of invertebrates such as the octopus, bees, and flies. The specifics of the visual reaction, not surprisingly, appear to be somewhat different.

Vertebrate eyes are endowed with many additional features designed for the increased effectiveness of vision for the needs of the particular animal concerned. These include such features as lids for protection, a cornea to help correct for the refractive differences between air (or water) and the eye tissue, a light-regulating iris diaphragm (also acting to increase acuity by reducing the aperture when possible), a focusing lens, and sometimes a reflective layer behind the retina to give the rod a second chance at the light quanta under very dark conditions. In addition, vertebrates often have a third (*parietal* or *pineal*) eye located as a projection from the roof of the forebrain. This parietal eye is still poorly understood but probably often handles the matter of assessing light intensity and day length for vegetative purposes and is not an image-forming eye used for vision in man's usual sense. Where the parietal eye is missing, however, some other parts of the nervous system—perhaps the retina, perhaps other parts—may take over these poorly delineated and poorly understood functions. The central nervous systems of vertebrates from fish to man are frequently conspicuously shielded from light by pigmented sheaths of connective tissue, reflective silvery layers, or thick hair, skulls, and so on. Is such shielding to protect the nervous system from damage by light or to prevent confusing light cues from being coded by the nervous tissue? In many animals, principally invertebrates, portions of the central nervous system that are light-exposed can be shown to function in special and often limited ways as photosensors. Eyes, often many per animal, as well as photosensitive nervous tissue areas or spots have evolved widely among the invertebrate phyla, but a discussion of these must be left to other sources.

Among vertebrates, our understanding of the kinds of things observed and measured by eyes has been inhibited by our self-awareness of what man can do visually. Man has, until recently, tended to conceive of a sensory world as bounded by his own abilities. Thus we early recognized that vision could report images, movement, color, relative light intensity, relative depth, relative direction, and distance. More recent evidence suggests that there are other kinds of specializations subserved by some vertebrate eyes, including the measurement of day length, the measurement of planes of polarization, the measurement of the position and movement of the sun and constellations (for navigation), the detection of movement of special or unique shapes and/or sizes of objects, and the determination of absolute direction. We had best leave this subject, adding only that understanding of many of these phenomena will be detailed soon; to conceive of others and to frame the experiments required to reveal and quantify their nature presents a particular intellectual challenge. Let us just mention three challenging clues. In many reptiles and birds, a fleshy, well-vascularized appendage, the *pecten*, projects from the retina into the vitreous humor of the eyeball. It contains no light sensitive cells; it forms a blind area. For what does it function? Over 30 speculative answers had already been formulated some 15 years ago. The answer is not yet in. Again, in birds and reptiles small "filters," consisting of colored oil droplets, lie in the light path of many visual cells. These droplets may be clear, yellow, orange, or red. What purpose do they serve? The retinas of some flying foxes, large, fruit-eating, visually orienting bats, are thrown into deep ridges or folds so that the visual cells are not in one hemispherical "plane" but are in very complex geometric relationship to each other. Why; to what advantage?

Eyes and ears have been relatively well studied, but even here our understanding is only beginning to jell. Studies of such other vertebrate senses as taste and olfaction have only barely begun to cross the threshold of sophistication. There is, as yet, no satisfactory formula for describing smell, for example. Some seven unrelated groups of fish have evolved systems for emitting electric current into the water surrounding them. Only recently has the function of these systems been probed. Only in the last few years has evidence identified the sense organs involved in monitoring this current or in utilizing it as a sensory system. (See Chapter 6 .)

High sensitivity is only one way in which receptor cells and their sensory neurons are specialized to provide useful information to the central nervous system. The customary interplay of excitation and inhibition occurs not only at synapses in the brain, where sensory axons have their endings, but may also occur in the sense organs themselves—especially in the retina of a vertebrate animal, which contains both receptors and a whole network of neurons and synapses through which impulses must pass before reaching the optic nerve. Stimulation of one visual receptor often elicits impulses that tend to inhibit synapses receiving impulses from adjacent ones. This competition between impulses arriving over axons from neighboring receptors heightens the discriminations that the retina can make between fine gradations of light and shade. The sensitivity of some sense organs can apparently be modulated by nerve impulses arriving from the central nervous system. The axons carrying such impulses synapse with the sense organs themselves. Very little as yet is known of how such centripetal motor outputs affect sensory function, but the story as it unfolds promises to be most exciting.

## THE ESSENCE OF A LIVING ANIMAL

Sense organs are not the only sources of nerve impulses for the central nervous system. Many of the cells also exhibit spontaneous activity, sending impulses down their axons even when isolated from all outside stimulation. There is thus no dearth of excitation within a brain; the question is, how does it all add up to a system that controls the animal's activities in an appropriately coordinated fashion? Of all the organ systems, the central nervous system contributes the most to making an animal the uniquely organized entity it is. Yet when we view this organ under the microscope, we see an endlessly tangled forest, where the cell bodies of the neurons, their branching axons and dendrites, and the glial cells are intermingled like a thicket of tree trunks, branches, and vines in what may appear at first to be utter confusion. Out of seeming chaos comes orderly regulation and control of all the varied and versatile organs of the animal's body.

Any part of any central nervous system differs from other organs in that each individual cell tends to have a specific function, which is often quite different from that of its immediate neighbor. Adjacent muscle fibers all contract in nearly the same way, but one of two adjacent neurons may excite a particular synapse, whereas the other causes inhibition. In an area of the brain where axons from the retina converge, one cell may respond when the light is turned on and another when it goes off; a third cell may ignore the light that flickers on and off but respond with a barrage of impulses when a spot of light moves across the surface of the retina. Furthermore, among the billions of neurons in a brain, there is not only a tremendous division of labor, but also a great duplication and overlap of function, so that many may be destroyed without appreciable damage to the effectiveness of the whole, organized brain.

In studying the workings of brains, biologists are just beginning to see the forest and the trees. The attempts to do so are actively pursued on many fronts—from highly detailed studies on submicroscopic structure and the biochemistry of the synaptic membranes where decisive events take place, to attempts to analyze the behavior of whole animals and achieve some orderly understanding of the factors that control what they do with the body machinery at their disposal. At this stage in the history of science, no one can explain how a brain really works, or even the nerve net of a coelenterate. This chapter has outlined what contemporary biologists believe are some of the most significant questions we can ask about nervous systems. With these in mind, one can hope to proceed with some perspective to study and achieve better understanding of animals and their behavior.

## FURTHER READING

Andrew, R. J., "The Origins of Facial Expressions," *Scientific American*, 213 (4): 88–94, 1965.

Aschoff, J., "Comparative Physiology: Diurnal Rhythms," *Annual Review of Physiology*, 25: 581–600, 1963.

———, (ed.), *Circadian Clocks.* Amsterdam: North-Holland Publishing Co., 1965.

Baker, P. F., "The Nerve Axon," *Scientific American,* 214 (3): 74–82, 1966.

Barnes, C. D., and C. Kircher, *Readings in Neurophysiology.* New York: Wiley, 1968.

Barrington, E. J. W., *An Introduction to General and Comparative Endocrinology.* New York: Clarendon Press, 1963.

Beck, S. D., *Animal Photoperiodism.* New York: Holt, Rinehart and Winston, 1963.

———, *Insect Photoperiodism.* New York: Academic Press, 1966.

Bekesy, G. von, *Sensory Inhibition.* Princeton: Princeton University Press, 1967.

Bentley, P. J., "Neurohypophyseal Function in Amphibians, Reptiles, and Birds," *Symposium of the Zoological Society of London,* 9: 141–152, 1963.

Brewer, C. V., *The Organization of the Central Nervous System.* London: Heinemann, 1961.

Bullock, T. H., and G. A. Horridge, *Structure and Function in the Nervous System of Invertebrates.* 2 Vols. San Francisco: Freeman, 1965.

Bünning, E., *The Physiological Clock.* New York: Academic Press, 1964.

Busnel, R. G., (ed.), *Acoustic Behaviour in Animals.* New York: Elsevier, 1963.

Carlisle, D. B., and F. Knowles, *Endocrine Control in Crustaceans.* New York: Cambridge, 1959.

Carr, A., *So Excellent a Fishe.* New York: Natural History Press, 1967.

Carthy, J. D., *An Introduction to the Behaviour of Invertebrates.* London: Allen, 1958.

Cohen, A. I., "Vertebrate Retinal Cells and Their Organization," *Biological Reviews of the Cambridge Philosophical Society,* 38: 427–459, 1963.

Dethier, V. G., *To Know a Fly.* San Francisco: Holden-Day, 1962.

———, *The Physiology of Insect Senses.* New York: Wiley, 1963.

DeVore, I. (ed.), *Primate Behavior.* New York: Holt, Rinehart and Winston, 1965.

Dijkgraaf, S., "The Functioning and Significance of the Lateral-Line Organs," *Biological Reviews of the Cambridge Philosophical Society,* 38: 51–105, 1963.

Dowling, J., "Night Blindness," *Scientific American,* 215 (4): 78–84, 1966.

Eccles, J. C., "The Synapse," *Scientific American,* 212 (1): 56–66, 1965.

Emlen, J. E., and R. L. Penney, "The Navigation of Penguins," *Scientific American,* 215 (4): 104–113, 1966.

Etkin, W. (ed.), *Social Behavior and Organization among Vertebrates.* Chicago: University of Chicago Press, 1964.

Florey, E., *An Introduction to General and Comparative Animal Physiology.* Philadelphia: Saunders, 1966.

Frisch, K. von, *Bees: Their Chemical Senses, Vision and Language.* Ithaca, N.Y.: Cornell University Press, 1950.

———, *The Dance Language and Orientation of Bees.* Cambridge, Mass.: Belknap, 1967.

Gorbman, A., and H. A. Bern, *A Textbook of Comparative Endocrinology*. New York: Wiley, 1962.

Griffin, D. R., *Listening in the Dark*. New Haven, Conn.: Yale University Press, 1958.

———, *Bird Migration*. New York: Natural History Press, 1964.

Harker, J. E., "The Physiology of Diurnal Rhythms," *Cambridge Monographs on Experimental Biology*, no. 13, 1964.

Hardy, J. D., "Physiology of Temperature Regulation," *Physiological Reviews*, 41: 521–606. 1961.

Heller, H., and R. B. Clark (eds.), "Neurosecretion," *Memoirs of the Society for Endocrinology*, 12: 1–455, 1962.

Hendricks, S. B., "How Light Interacts with Living Matter," *Scientific American*, 219 (3): 174–186, 1968.

Hoar, W., *General and Comparative Physiology*. Englewood Cliffs, N.J.: Prentice-Hall, 1966.

Holst, E. von, and U. von Saint-Paul, "Electrically Controlled Behavior," *Scientific American*, 206 (3): 50–59, 1962.

Horridge, G. A., *Interneurons*. San Francisco: Freeman, 1968.

———, "Integrative Action of the Nervous System," *Annual Review of Physiology*, 25: 523–544, 1963.

Hubbard, R., and A. Kropf, "Molecular Isomers in Vision," *Scientific American*, 216 (6): 64–76, 1967.

Hubel, D. H., "The Visual Cortex of the Brain," *Scientific American*, 209 (5): 54–62, 1963.

Jacobsen, M., and M. Beroza, "Insect Attractants," *Scientific American*, 211 (2): 20–27, 1964.

Jolly, A., *Lemur Behavior; a Madagascar Field Study*. Chicago: University of Chicago Press, 1966.

Katz, B., *Nerve, Muscle and Synapse*. New York: McGraw-Hill, 1966.

Kellogg, W. N., *Porpoises and Sonar*. Chicago: University of Chicago Press, 1961.

Lehrman, D. S., "The Reproductive Behavior of Ring Doves," *Scientific American*, 211 (5): 48–54, 1964.

Levey, R. H., "The Thymus Hormone," *Scientific American*, 211 (1): 66–77, 1964.

Lindauer, M., *Communication among Social Bees*. Cambridge, Mass.: Harvard University Press, 1961.

Lissmann, H. W., "Electric Location by Fishes," *Scientific American*, 208 (3): 50–59, 1963.

McCartney, W., *Olfaction and Odours*. New York: Springer-Verlag, 1968.

McElroy, W. D., and B. Glass (eds.), *Light and Life*. Baltimore: The Johns Hopkins Press, 1961.

Marler, P., and W. J. Hamilton III, *Mechanisms of Animal Behavior*. New York: Wiley, 1966.

Maturana, H. R., J. Y. Lettvin, W. S. McCullock, and W. H. Pitts, "Anatomy and Physiology of Vision in the Frog *(Rana pipiens)*," *Journal of General Physiology*, 43 (6, pt. 2): 129–175, 1960.

Miller, W. H., F. Ratcliff, and H. K. Hartline, "How Cells Receive Stimuli," *Scientific American*, 205 (3): 233–238, 1961.

Milne, L. J., and M. Milne, *The Senses of Animals and Men*. New York: Atheneum, 1962.

Moncrieff, R. W., *The Chemical Senses*. Cleveland: CRC Press, 1967.

Muntz, W. R. A., "Vision in Frogs," *Scientific American*, 210 (3): 111–119, 1964.

Mykytowycz, R., "Territorial Marking by Rabbits," *Scientific American*, 218 (5): 116–126, 1968.

Neisser, U., "The Processes of Vision," *Scientific American*, 219 (3): 204–214, 1968.

Pribam, K., "The Neurophysiology of Remembering," *Scientific American*, 220 (1): 73–86, 1969.

Roeder, K. D., *Nerve Cells and Insect Behavior*. Cambridge, Mass.: Harvard University Press, 1963.

———, "Moths and Ultrasound," *Scientific American*, 212 (4): 94–102, 1965.

Rosenblith, W. A. (ed.), *Sensory Communication*. Cambridge, Mass.: M.I.T. Press, 1961.

Schaller, G., *The Year of the Gorilla*. Chicago: University of Chicago Press, 1964.

———, *The Deer and the Tiger*. Chicago: University of Chicago Press, 1967.

Scharrer, E., and B. Scharrer, *Neuroendocrinology*. New York: Columbia University Press, 1963.

Stettner, L. J. and K. A. Matyniak, "The Brain of Birds," *Scientific American*, 218 (6): 64–76, 1968.

Sudd, J. H., *An Introduction to the Behavior of Ants*. New York: St. Martin's, 1967.

Thorpe, W. H., *Bird Song: the Biology of Vocal Communication and Expression in Birds*. New York: Cambridge, 1961.

Tinbergen, N., *Animal Behavior*. New York: Time, Inc., Book Division, 1965.

Turner, C. D., *General Endocrinology*, 4th ed. Philadelphia: Saunders, 1966.

Wells, M. J., *Brain and Behavior of Cephalopods*. London: Heinemann, 1962.

Wenner, A. M., "Sound Communication in Honeybees," *Scientific American*, 210 (4): 117–124, 1964.

Wilson, E. O., "Pheromones," *Scientific American*, 208 (5): 100–114, 1963.

———, "The Social Biology of Ants," *Annual Review of Entomology*, 8: 345–368, 1963.

———, and W. H. Bossert, "Chemical Communication among Animals," *Recent Progress in Hormone Research*, 19: 673–716, 1963.

Wilson, V. J., "Inhibition in the Central Nervous System," *Scientific American*, 215 (5): 102–110, 1966.

Wright, R. H., *The Science of Smell*. London: G. Allen, 1964.

Wurtman, R. J., and J. Axelrod, "The Pineal Gland," *Scientific American*, 213 (1): 50–60, 1965.

Young, W. C. (ed.), *Sex and Internal Secretions*, 3rd ed., vols. 1 and 2. Baltimore: Williams & Wilkins, 1961.

# *Index*

## A

N



O